THE USE AND FATE OF PESTICIDES IN VEGETABLE-BASED AGROECOSYSTEMS IN GHANA

T0186370

Promoter: Prof. dr. H. J. Gijzen
Professor of Environmental Biotechnology
UNESCO-IHE/Wageningen University
The Netherlands

Co-promoter: Dr. P. Kelderman
Senior lecturer in Environmental Chemistry
UNESCO-IHE Institute for Water Education
The Netherlands

Awarding Committee: Dr. P. Drechsel
International Water Management Institute (IWMI) Ghana

Prof. dr. P. N. L. Lens
UNESCO-IHE Institute for Water Education
The Netherlands

Dr. M. A. Siebel
UNESCO-IHE Institute for Water Education
The Netherlands

Prof. dr. ir. A. Veldkamp
Wageningen University
The Netherlands

Dr. M. Yangyuoru
University of Ghana
Ghana

The use and fate of pesticides in vegetable-based agroecosystems in Ghana

DISSERTATION

Submitted in fulfilment of the requirements of
the Academic Board of Wageningen University and
the Academic Board of the UNESCO-IHE Institute for Water Education
for the Degree of DOCTOR
to be defended in public
on Thursday, 17 January 2008 at 15:30 hours
in Delft, The Netherlands

by

WILLIAM JOSEPH NTOW
born in Kumasi, Ghana

Taylor & Francis is an imprint of the Taylor & Francis Group, an informa business.

© 2008, William Joseph Ntow

All rights reserved. No part of this publication or the information contained herein may be reproduced, stored in a retrieval system, or transmitted in any form or by any means, electronic, mechanical, by photocopying, recording or otherwise, without written prior permission from the publishers.

Although all care is taken to ensure integrity and the quality of this publication and the information herein, no responsibility is assumed by the publishers nor the author for any damage to the property or persons as a result of operation or use of this publication and/or the information contained herein.

Published by:
Taylor & Francis/Balkema
PO Box 447, 2300 AK Leiden, The Netherlands
e-mail: Pub.NL@tandf.co.uk
www.balkema.nl, www.taylorandfrancis.co.uk, www.crcpress.com

ISBN 978-0-415-46274-7 (Taylor & Francis Group)
ISBN 978-90-8504-836-7 (Wageningen University)

To the memory of Dr. Akosua Werekoa (aka Awo), this thesis is dedicated to her, my wife, Mrs Bernice Worlanyo Ntow and my parents, Francis Andrews Ntow and Margaret Kyei (Yaa Saah), whom God has used to bring me this far.

Contents

Acknowledgements

I offer my sincere thanks to every staff member of the Council for Scientific and Industrial Research - Water Research Institute (CSIR WRI), Accra, the International Water Management Institute (IWMI), Ghana Office, Accra, and the UNESCO-IHE Institute for Water Education, Delft, along with Benjamin Osei Botwe, Jonathan Ameyibor, Laud Mike Tagoe, friends and colleagues from Kinneret Limnological Laboratory, Migdal, Israel, who had a hand in producing this thesis.

Firstly, I acknowledge the financial support given to me by the Dutch Government (through the UNESCO-IHE Institute for Water Education), the International Water Management Institute, the International Foundation for Science/Organisation for the Prohibition of Chemical Weapons, and the Ghana Government, without which this study would not have been possible.

I am grateful to Prof. dr. Huub J. Gijzen (Promoter), Dr. Peter Kelderman (Co-promoter) and Dr. Pay Drechsel (Supervisor) who worked closely with me throughout the period of the study. Their constructive criticisms, comments and advice helped to shape this thesis. They also provided encouragement and support to me during the study. They deserve my sincere thanks.

I am also grateful to my parents, Francis Andrews Ntow, Margaret Kyei Ntow, my brothers and sisters (Yaw, Grace, Joe, Kwabena, Afua, Ama and Francis), and my Uncle Kyei Baffour for their prayers, encouragement and moral support.

All my efforts would be impossible without strong spiritual and moral backing of my beloved wife, Bernice Worlanyo Ntow. She was a strong source of motivation, and provided secretarial assistance to me during the entire study. My children, Nana Kwasi Ntow, Ama Saah Ntow and Kwabena Pobi Ntow were sources of moral support and encouragement to me. These cherished people deserve my sincere thanks.

I am indebted to my friends in the Netherlands, Mr/Mrs Boakye Yiadom (Delft), Elder/Mrs Bekoe (Delft), Martha (Den Haag), Mary (Den Haag), Apostle Ouadrago and the entire congregation of Church of Pentecost (Den Haag) who took care of my physical and spiritual need while in the Netherlands. Also these people deserve my sincere thanks.

Finally, I give Glory to God Almighty through Jesus Christ, my Lord and Saviour, who saw me through the entire PhD programme. I owe my very life, and for that matter this thesis, to God. In His own time, He makes things beautiful. God, I am grateful to you.

Chapter 1

Introduction

Introduction

General

Worldwide pesticide usage has increased tremendously since the 1960s. It has largely been responsible for the "green revolution", *i.e.* the massive increase in food production obtained from the same surface of land with the help of mineral fertilisers (nitrogen, phosphorus, potassium), more efficient machinery and intensive irrigation. The use of pesticides helped to significantly reduce crop losses and to improve the yield of crops such as corn, maize, vegetables, potatoes and cotton. Notwithstanding the beneficial effects of pesticides, their adverse effects on environmental quality and human health have been well documented worldwide and constitute a major issue that gives rise to concerns at local, regional, national and global scales (Planas *et al.*, 1997; Huber *et al.*, 2000; Kidd *et al.*, 2001; Ntow, 2001; Cerejeira *et al.*, 2003). Residues of pesticides contaminate soils and water, persist in the crops, enter the food chain, and finally are ingested by humans with foodstuffs and water. Furthermore, pesticides can be held responsible for contributing to biodiversity losses and deterioration of natural habitats (Sattler *et al.*, 2006). There have been reported instances of pest resurgence, development of resistance to pesticides, secondary pest outbreaks and destruction of non-target species. Despite the fact that pesticides are also applied in other sectors, agriculture can undoubtedly be seen as the most important source of adverse effects (Hoyer and Kratz, 2001; cited in Sattler *et al.*, 2006).

The application of different pesticides varies with the region. For instance, in North America and Western Europe, due to high costs of labour, the chemical control of weeds with herbicides is much more common than in East Asia and Latin America. However, in many tropical regions with insect pests and plant diseases, insecticides are also applied in large amounts, both in small farms and in industrial plantations. Regarding the current use of pesticides: while developed countries such as USA, Canada and countries in the European Union proceed in a direction of fewer chemicals and more "green products", developing countries often follow a different direction in these matters. There is a large need here for an increase in agricultural production, and the use of crop protection chemicals seems a simple way for obtaining better crop yields. This holds in particular for the country of this study, *viz.* Ghana, West Africa, a predominantly agricultural country.

Problem definition

Ghana (5° 36' N, 0° 10' E), officially the Republic of Ghana, is a country in Africa bordering Cote d'Ivoire to the west, Burkina Faso to the north, Togo to the east, and the Gulf of Guinea to the south (Figure 1.1).
Ghana has roughly the size of the state of Britain. The coastline is mostly a low, sandy shore supported by plains and tracts of stunted vegetation and intersected by several rivers and streams. The geology of Ghana is dominated by metavolcanic Paleoproterozoic Birimian sequences and the clastic Tarkwaian (http://www.uoguelph.ca/~geology/rocks_for_crops/28ghana.PDF; accessed March 23, 2007).

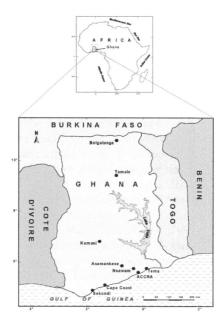

Figure 1.1.   Map of Ghana, its boundaries and Lake Volta. Also shown are some major towns

A tropical rain forest belt, broken by heavily forested hills and many streams and rivers, extends northward from the shore. North of this belt, the land is covered by low bush, park-like savanna and grassy plains. Ghana's climate is tropical; the eastern coastal belt is warm and comparatively dry, whereas the southwest corner is hot and humid, and the north is hot and dry. Lake Volta (see Figure 1.1), the world's largest artificial lake, extends through large portions of eastern Ghana. Ghana is divided into 10 regions, which are then subdivided into a total of 138 districts. Accra is the capital and largest city. Table 1.1 presents some "quick facts" on Ghana.

Agriculture is Ghana's most important economic sector, employing about 60% of the national work force, mainly small landholders on a formal and informal basis (Gerken *et al.*, 2001) and accounting for about 40% of total GDP and export earnings (see Table 1.1). The climatic zones range from dry savanna to wet forest and run in east-west bands across the country. Agricultural crops including yams, grains, cocoa, oil palms, kola nuts and timber, form the base of Ghana's economy (http://countrystudies.us/ghana/77.htm ; accessed March 15, 2007). Compared to this, the contribution of the more traditional vegetables such as okra, pepper, tomato, onion and eggplants to the agricultural GDP is low. However, considering their contribution to agricultural GDP alongside the size of land area devoted to the cultivation of all crops, the traditional vegetables turn out to be better performers. For example, the share in total area cultivated and value for the vegetables (tomato, okra, pepper, and eggplant) is 3 and 7%, respectively, while for cereals (maize, millet, sorghum and rice), these values are 35 and 8%, respectively (Table 1.2).

Table 1.1. Ghana: Quick Facts

	Year 2005
Population, total (millions)	22.1
Urban population (% of total population)	48
Population growth (annual %)	2
Age structure (% of population):	
0-14 years	39
15-64 years	58
65 years and over	3
Surface area (km^2) (thousands)	238.5
Average annual temperature ($^\circ$C)	30
Rainfall (annual, mm)	1,015-2,300
Life expectancy at birth, total (years)	57.2
Mortality rate, infant (per 1,000 live births)	68.0
Child malnutrition (% of children under 5 years)	22
Access to an improved water source (% of population)	75
Literacy (% of population, ages 15 and above)	58
Gross Domestic Product, GDP (US$ billions)	10.7
GDP (average annual growth)	5.8
GDP per capita, US$	486
Agriculture, value added to GDP (%)	39

Sources: World Develpment Indicators database, April 2006
          https://cia.gov/cia/publications/factbook/print/gh.html; accessed March 25, 2007

Table 1.2. Growth rates and contributions of different crops to Agricultural GDP

Crops	Share in value (%)	Share in area (%)	Rate of growth (%)	
			Area	Yield
Maize	4.80	17.0	2.00	3.00
Rice	1.00	2.23	8.00	2.00
Millet	1.30	7.27	2.00	-
Sorghum	1.24	8.41	2.00	1.50
Cassava	19.2	12.1	2.00	3.00
Cocoyam	10.2	6.08	2.00	-
Yam	16.7	6.32	2.36	0.75
Plantain	13.0	5.32	2.00	-
Groundnut	3.83	4.66	2.36	3.00
Tomato	2.79	0.61	1.00	2.00
Okra	3.16	0.85	3.00	2.00
Pepper	0.14	1.71	3.00	2.00
Eggplant	0.56	0.07	3.00	2.00
Beans	15.3	5.12	1.00	2.00
Cocoa	0.92	18.8	5.00	2.00
Oil palm	0.07	1.58	2.00	2.00
Rubber	0.06	0.06	4.00	2.00
Coffee	0.16	0.02	4.00	2.00
Cotton	0.12	0.32	8.00	2.00
Tobacco	3.47	0.06	8.00	2.00
Orange	0.09	0.55	3.00	2.00
Pineapple	0.09	0.06	3.00	2.00
Total crops/growth rates	98.2%	99.2%	3.31%	1.58%

Source: Nurah, 1999

Within the agricultural sector, vegetable production plays an important and varied nutritional as well as socio-economic role. Vegetable production in Ghana has developed from a mainly subsistence activity carried out by women to a commercial activity carried out by mainly young men and women. Vegetable production is done in rural, peri-urban and urban areas. Tomato, eggplant, pepper and onion are grown in all ten regions of Ghana. However, some regions are more efficient and specialised in the production of only one or two out of above four crops. One of the biggest problems confronting vegetable farmers in Ghana is disease and pests which ravage their crops. Vegetables, generally, attract a wide range of pests and diseases, and can require intensive pest management (Dinham, 2003). The pest control practices in vegetable production in Ghana involve applications of highly toxic pesticides which are most of the time misapplied, and which result in pesticide contamination of the produce itself as well as the environment. While Ghana's elite is becoming increasingly concerned about the adverse long-term effects of pesticides on the environment and the health of the country's resources, little scientific research has been done to address the issue.

Within the context of efforts to achieve safe, sound and sustainable production of vegetables, safe pesticide management plays a crucial role. Pesticide management includes all aspects of the safe, efficient and economic use and handling of pesticides (Bull, 1982). The proper use of pesticides in Ghana means taking into account the health, social and economic realities of life. It implies using pesticides which can safely be applied, and only then when necessary, in the appropriate health, social and environmental context.

The present study

This study focuses on the analysis of data on the types and dosages of pesticides applied as well as their use, and practices of use, in vegetable production in different areas in Ghana. The study uses field experiments to provide insight into the overall fate of pesticides on vegetable plots. By using the field trials in order to understand the fate of the pesticides on agricultural plots, a means will be accessible to establish the fate of the pesticides in the wider environment. The persistence of the pesticides, their extent of pollution and occurring trends (spatial and seasonal), in various compartments of the vegetable field ecosystem (on-farm and off-farm) will be assessed. The fate of the pesticides in the environment will be evaluated mainly by laboratory measurements on levels of residues of a selected pesticide. Samples were analysed by state-of-the-art technology, *viz.* Gas Chromatography coupled with Mass Spectrometry (GC-MS). Simultaneously, measurement were carried out on residue levels in farmers' breast milk and blood, as well as cholinesterase activity in blood. These cholinesterase tests provide epidemiological evidence of exposure to the chemicals. Farmers' activities and practices that lead to exposure to the chemicals were investigated to provide support to the laboratory tests.

The study is based on the null hypothesis that the incidence and magnitude of environmental and public health effects of pesticide use in agriculture is independent of type, use (frequencies and dosages), and persistence of pesticides.

Objectives

The specific objectives of this study are to:
- Review the current patterns and practices of use of pesticides in Ghanaian vegetable farming
- Establish the distribution, persistence and dissipation of endosulfan in a tomato growing field
- Establish the levels of historical and current-use pesticides in environmental, crop and human fluid samples in selected areas of vegetable production in Ghana
- Assess the public health risks due to pesticide exposure in the region .

Outline of the thesis

The Thesis starts with an introduction (this Chapter) that presents general information on Ghana and its agricultural crops. The research outline is given and aim, scope, strategy and objectives of the research are presented. In Chapters 2-7, the research carried out and its pertinent findings are described in detail.

How are pesticides used by farmers in vegetable cultivation (*i.e.* what are the types of chemicals that a farmer uses; to what extent does a farmer rely on these chemicals and what is his/her timing of sprays?); what is the farmer's level of knowledge of pesticides; what is his/her perception of the chemicals potential for harm, and what is the farmer's attitude towards personal protective clothing? These issues will be discussed in Chapter 2 of this Thesis.

The research described in Chapter 3 used endosulfan applied to field-grown tomato to show the pesticide's behaviour, persistence and mobility under field conditions; the selection of endosulfan and the vegetable crop was based on information obtained in Chapter 2. The research described in Chapter 3 will also provide insight into the overall fate of pesticides on agricultural plots.

By using the results of above field studies, a means will be accessible to establish the fate of pesticides in the wider environment. The persistence of pesticides, their extent of pollution and occurring trends (spatial and seasonal) in various compartments of the ecosystem: water, water-bed sediment, as well as human exposure (blood enzyme activity, residues of persistent compounds in blood and mothers' breast milk, and dietary intake) will be assessed in Chapters 4-7.

Chapter 4 discusses the monitoring results for the off-site effects of pesticides use in vegetable cultivation. The chapter deals with the pollution (extent and trend) of current-use pesticides in water and water-bed sediment. Pesticide runoff via the drainage water from the vegetable field into two small streams draining the field, will be considered here.

In Chapter 5, cholinesterase activity in blood of exposed farmers will be discussed as an epidemiological evidence of health effect. The results of investigations into residue levels of persistent organochlorine pesticides in blood serum and breast milk of exposed farmers are the focus of Chapter 6; the Chapter takes a detailed look at the human cost of inappropriate and ill-advised pesticide use.

The presence of pesticides in food crops (vegetable crops on market tables) as a possible source of exposure, and their potential as possible carcinogens, are assessed and discussed in Chapter 7.

Finally, in Chapter 8 of this Thesis the following issues will be presented: overall summary of the Thesis and conclusions of this research, recommendations for future research and concluding remarks.

References

Bull, D. (1982) A growing problem: Pesticides and the Third World poor, OXFAM, Oxford, 192 pp. Carvalho, F.P. (2006) Agriculture, pesticides, food security and food safety. *Environmental Science & Policy* 9:685-692.

Cerejeira, M.J., Viana, P., Batista, S., Pereira, T., Silva, E., Valerio, M.J., Silva, A., Ferreira, M., and Silva-Fernandes, A.M. (2003) Pesticides in Portuguese surface and ground waters. *Water Research* 37:1055-1063.

Dinham, B. (2003) Growing vegetables in developing countries for local urban populations and export markets: problems confronting small-scale producers. *Pest Manag Sci* 59:575-582.

Gerken, A., Suglo, J.V. and Braun, M. (2001) Pesticide policy in Ghana. MoFA/PPRSD, ICP Project, Pesticide Policy Project/GTZ, Accra, Ghana, 185 pp.

Huber, A., Bach, M., Frede, H.G. (2000) Pollution of surface waters with pesticides in Germany: modeling non-point source inputs. *Agriculture Ecosystems and Environment* 80:191-204.

Kidd, K.A., Bootsma, H.A. and Hesslein, R.H. (2001) Biomagnification of DDT through the benthic and pelagic food webs of Lake Malawi, East Africa: Importance of trophic level and carbon source. *Environmental Science and Technology* 35:14-20.

Ntow, W.J. (2001) Organochlorine pesticides in water, sediment, crops and human fluids in a farming community in Ghana. *Archives of Environmental Contamination and Toxicology* 40:557-563.

Nurah, G.K. (1999) A baseline study of vegetable production in Ghana. National Agricultural Research Project (NARP) Report, Accra, 151 pp.

Planas, C., Caixach, J., Santos, F.J., and Rivera, J. (1997) Occurrence of pesticides in Spanish surface waters. Analysis by high-resolution gas chromatography coupled to mass spectrometry. *Chemosphere* 34:2393-2406.

Sattler, C., Kächele, H., and Verch, G. (2006) Assessing the intensity of pesticide use in agriculture. *Agriculture, Ecosystems and Environment* doi:10.1016/j.agee.2006.07.017.

Chapter 2

Farmer perceptions and pesticide use practices in vegetable production in Ghana

Publication based on this chapter:

William J Ntow, Huub J Gijzen, Peter Kelderman and Pay Drechsel (2006). Farmer perceptions and pesticide use practices in vegetable production in Ghana. *Pest Manag Sci.* 62:356-365.

Farmer perceptions and pesticide use practices in vegetable production in Ghana

Abstract

As an initial part of a programme aimed at promoting safe and sound agricultural practices in Ghana, a study was made of farmers' perceptions of pesticides for use and application in vegetable production, using a small survey of 137 farmers who applied pesticides. Field surveys, interviews, questionnaires and analytical games were used to obtain information on the type, scope and extent of use of pesticides, farmers' knowledge of pesticides, and their perceptions about the chemicals' potential for harm. Data from this sample of farmers were used to describe the status of use of pesticides in vegetable cultivation in Ghana. Using χ^2 tests, associations between farmers' age and possible pesticide poisoning symptoms, their farm size and method of spraying pesticides, and their perception of pesticide hazard and its perceived effectiveness against pests were also examined. The survey showed that knapsack sprayers were the most widely used type of equipment for spraying pesticides. However, on large-scale vegetable farms of 6-10 acres, motorised sprayers were also used. Various inappropriate practices in the handling and use of pesticides caused possible poisoning symptoms among those farmers who generally did not wear protective clothing. Younger farmers (<45 years of age) were the most vulnerable group, probably because they did more spraying than older farmers (>45 years of age). Farmers did not necessarily associate hazardous pesticides with better pest control. The introduction of well-targeted training programmes for farmers on the need for and safe use of pesticides is advocated.

Keywords: insecticide; pest control; pesticides; pest management; vegetables

Introduction

Urban food needs in cities and towns in Ghana are growing, and increasingly vegetables are grown in urban and peri-urban areas to meet the demand. However, traditional vegetable farming systems (i.e. without any chemical input) are incapable of meeting this challenging demand. For instance, pests and diseases, which pose big problems in vegetable production, require intensive pest management to control them. Chemical pesticide use is a common practice to control pests and diseases in vegetable cultivation in Ghana. However, besides their beneficial effects, pesticides are accepted as having potential environmental and public health impacts as well. If improperly used, pesticides can cause direct human poisoning, accumulate as residues in food and the environment or lead to the development of resistant strains of pests. These problems can arise from misuse of the pesticides or over-reliance on them, particularly if the users are not aware of these potential problems. In Ghana there are already some levels of contamination of pesticides in water, sediment, crops and human fluids in areas of highly intensive vegetable production (Ntow, 2001). There also exist species of aphids, which have developed resistance to some insecticides, and there are probably other pests resistant to other pesticides, which are as yet undetected.

While pesticides are generally considered a panacea for farmers' pest concerns, farmers' perceptions and use of the chemicals have not received much attention. In Ghana there has not been any known comprehensive study. However, the perceptions of the farmers regarding, in particular, pesticide risks to human health are important for a number of reasons: (Warburton *et al.*, 1995) first, they may influence decisions regarding pesticide use; second, if these perceptions differ from expert opinion, it is useful to know why and whether they lead farmers to take more risks than they realise; third, they may influence the methods of protection used against pesticides; and, last, technical advice given to farmers on pesticide use and crop protection may be inappropriate and irrelevant if it does not tally with their own views of pesticide health effects.

The study had the following objectives.

- To determine the extent of use of pesticides, i.e. the types of chemicals that farmers use, pesticide use practices, to what extent the farmers rely on the chemicals, and the timing of spraying.
- To assess the extent of farmers' knowledge of pesticides and their perceptions about the chemicals' effectiveness and potential for harm.
- To examine farmers' pesticide management practices.

Materials and methods

Diagnostic surveys, formal and informal interviews, field observations and analytical games were used to gather information about farmers' pesticide use, perceptions on health effects, and pest management practices. Formal interviews dealt with the general vegetable farming system and farmers' pest control practices, while the other methods assessed farmers' pesticide use and perception of pesticide hazards.

In the formal interviews, structured questionnaires were used to collect information from the farmers. Four groups of data were asked for: (1) personal, (2) lifestyle, i.e. hygiene, eating, smoking and drinking habits, (3) farm details and work history and (4) pesticide use practices and management (Table 2.1). Some of the questions included in the questionnaire are not relevant to the present chapter. All questions were closed questions in a multiple-choice format, so that respondents had only to tick the appropriate answer. Some questions demanded multiple answers. External peer reviewers and collaborators at the UNESCO-IHE Institute for Water Education reviewed the questionnaire. The author administered the questionnaires at various locations, including farmers' homes, farms and school classrooms. In all cases the farmers were notified of the impending interview through their chiefs, group leaders or fellow farmers. The study objectives were explained to the chiefs, group leaders and the farmers in the identified areas, and their consent to participate in the study was obtained. The principal investigator translated the questionnaires into local and easily understandable languages, taking to retain their original meaning. In some instances the principal investigator sought assistance from other farmers to translate the questionnaires into the local language. Some of the interviews were recorded on video. However, pen and paper were used to record all interviews. Other information about the farmers and their farms was obtained that enabled the location of the farms for field observations. In addition to the interviews, field observation surveys and spraying practices of respondent farmers were discreetly conducted. The farmers were not informed beforehand in order to avoid modifications in pesticide handling behaviour and to reduce 'interviewer/respondent' bias.

Data were double-keyed for quality control. Cases with missing values were not included in the analysis. The questionnaire data were analysed by the statistical software SPSS (release 11.0; SPSS Inc., Chicago, IL, USA). First we summarised responses over all populations (we calculated frequencies for the responses). We also calculated means for age, sex, marital status, years of farming and acres of plot owned, and means and standard deviations for age and years of farming. Then we

Table 2.1. Overview of questions from questionnaire concerning pesticide use that was given to   egetable farmers

Data group	Description
Personal data	Sex; marital status; age; education
Lifestyle	Do you smoke cigarettes (yes/no); do you drink (local gin, beer/Guinness, millet beer); how many times in a day do you eat (one, two, three); how often do you take your bath (once, twice, thrice in a day); what time of day do you take your bath; do you wash your hands before eating (yes/no); if you wash your hands before eating, with what do you wash (only water, water and soap, other)?
Work history and farm details	Years total worked as a farmer; occupation in past 12 months; size of farm (acres, % under vegetable cultivation); type of vegetables; cropping system
Pesticide use and management	Crops and their pesticides; spraying frequency; why do you apply pesticides at a particular time; method of application; knowledge of application rate; direction of spraying; type of protective clothing; re-entry period; storage place for pesticides; how do you dispose of empty pesticide containers; do you own a sprayer (yes/no); do you wash or clean sprayer after spraying; how do you dispose of the water you use to wash sprayer; what symptoms or poisoning cases have you experienced after a spray event?

compared the distribution of possible pesticide poisoning symptoms for different age categories (i.e. younger (<45 years) and older (>45 yeas) farmers, hypothesising that age appears to increase the proportion of farmers reporting possible pesticide poisoning symptoms). We used the χ^2 test to investigate a possible association. The independence of method of pesticide application and farm size was also assessed by the χ^2 test. We divided farm size into four categories: subsistence = <1 acre, small-scale = 1-5 acres, medium-scale = 6-10 acres and large scale = >10 acres.

Group meetings were also held to gather information simultaneously from a large number of farmers. However, care was taken in generalising from responses given at the group meetings, because certain categories of farmers were under-represented at the meetings (e.g. women). At the group meetings the investigators gathered information on cropping patterns, planting dates, vegetables grown, farming practices and input use.

Pesticide perceptions and health risks were assessed through a ranking game described by Warburton *et al.* (1995). Farmers, labourers and their spouses participated in the game. All participants were individually shown a total of 33 empty containers and/or labels of pesticides commonly available in the area. The containers and/or labels were shown one by one, and the names of the pesticides were read out to ensure that each participant knew what it was. They were asked which ones they recognised (but not necessarily used); the unfamiliar pesticides were removed and noted in the questionnaire. From the familiar containers and/or labels the respondents were asked which ones were thought to be generally effective in controlling pests, hence dividing the containers and/or labels into two piles: 'effective' and 'ineffective'. The 'ineffective' pile was removed and the 'effective' pesticides were ranked accordingly. Once the pesticides were ranked, respondents were asked to describe how effective each pesticide was on a scale of 1-5 (Table

2.2). Finally they were asked their reasons for ranking a particular pesticide as the most effective. A similar process was followed to rank the pesticides according to hazard: participants were asked to select and rank empty containers and/or labels of pesticides thought to be hazardous. Again they were asked to describe the pesticides on a scale of 1-5 (Table 2.2) and to provide reasons for ranking a particular pesticide as the most hazardous. The relationship between farmers' perception of pesticide hazard and pesticides' perceived effectiveness against pests was assessed by the χ^2 test.

Results

Table 2.3 gives the demographics for the 137 respondents. The average age was 45 years, 117 (85.4%) respondents were male and 111 (81.0%) farmers were married and living with their wives. On the average, each farmer had worked on a farm for at least 20 years and applied pesticides themselves on their own farms, with the majority (76.6%) of the farmers owning between 1 and 5 acres of plot.

Table 2.2. Description of ranking levels used for pesticide ranking game

Effectiveness	
1	Very effective: 75-100% insects killed
2	Effective: 50-75% insects killed
3	Small effect only: < 50% insects killed
4	No effect
5	Makes the insect problem worse
Hazard	
1	Extremely hazardous: likelihood of hospitalisation or long-term illness
2	Moderately hazardous: likelihood of more than 2 days sick and need to see a doctor
3	Slightly hazardous: likelihood of dizziness or vomiting or blurred vision or skin sores
4	Least hazardous: likelihood of some dizziness, tiredness, or headache
5	No effects

Table 2.3. Demographic characteristics of 137 vegetable farmers

Variable	Mean (±SD) (range) or %
Age (years)	44.9 (± 10.9) (24-80)
Sex (male)	85.4%
Marital status (married)	81.0%
Years of farming (worked as a farmer)	21.2 (± 10.5) (2-60)
Acres owned (1-5)	76.6%

Pesticides

A total of 43 pesticides were found in use in vegetable farming in Ghana. This figure was obtained as a direct summation of pesticides applied on farms, but it could be lower than the actual number of pesticides in use. The pesticides comprised insecticides, fungicides and herbicides. Herbicides (44%) were the class of pesticides most used in vegetable farming in the areas surveyed, followed by insecticides (33%) and fungicides (23%).   In Table 2.4 the classification of these pesticides by the type of pests they control, active ingredient, chemical group and WHO Hazard Category is presented. The herbicides and fungicides used are mostly under WHO Hazard Category III, with a few under Hazard Category II. All the insecticides used are under Hazard Category II, which WHO classifies as moderately hazardous.   This category includes organochlorines (OCs), organophosphates (OPs) and pyrethroids. Endosulfan was the only OC mentioned in use in the survey.

Table 2.4. Types of pesticides applied in vegetable production

Pesticide type (% of total number in use)	Active ingredient (AI)	Chemical group	Chemical AI Hazard Category (WHO)	Registered for use on
Herbicide (44%)	Pendimethalin	Dinitroaniline	III	Tomatoes, onions
	2,4-D	Aryloxyalkanoic acid	II	Rice, sugarcane
	propanil	Anilide	III	Rice
	MCPA-thioethyl	Aryloxyalkanoic acid	III	Not registered
	Oxadiazon	Oxadiazole	III	Not registered
	Oxyfluorfen	Diphenyl ether	III	Not registered
	Bensulfuron-methyl	Sulfonylurea	III	Rice
	Glyphosate	Glycine derivative	III	Various crops
	Paraquat dichloride	Bipyridylium	II	Various crops
	Acifluorfen	Diphenyl ether	III	Not registered
	Metolachlor	Chloroacetamide	III	Not registered
	Phenmedipham	Carbamate	III	Not registered
	Mancozeb	Carbamate	III	Mangoes, vegetables
Fungicide (23%)	Metalaxyl-M	Acylalanine	II	Not registered
	Thiophanate-methyl	Benzimidazole	III	Various crops
	Carbendazim	Benzimidazole	III	Not registered
	Benomyl	Benzimidazole	III	Not registered
	Lambda-cyhalothrin	Pyrethroid	II	Vegetables
	Chlorpyrifos	Organophosphorus	II	Citrus, public health
Insecticide (33%)	Endosulan	Organochlorine	II	Cotton
	Dimethoate	Organophosphorus	II	Not registered
	Cypermethrin	Pyrethroid	II	Not registered
	Deltamethrin	Pyrethroid	II	Various crops

Pesticides use and management

Table 2.5 shows the frequency distribution (count and percentage) of the responses pooled over all populations of vegetable farmers in the survey. A higher proportion of farmers had only basic education (up to middle school), cited the presence of pest as a major criterion for pesticide application, wore no or partial protective clothing and used a knapsack sprayer to spray pesticide in the wind direction. Again, a higher proportion of farmers returned to their farm within 48 h after spraying pesticide and disposed of their sprayer wash water and empty pesticide containers on the farm. The most frequently reported possible pesticide poisoning symptoms were body weakness (36.7%) and headache/dizziness (31.0%).

However, opinions differed among groups regarding the timing of spraying; here we present a few significant results only. For respondents in Ashanti Region (AR;

Figure 2.1), calendar spraying was considered the most important factor (about 60% of farmers responded). We summarise in Table 2.6 the spraying records of representative farmers at Akumadan (a major vegetable growing centre in Ashanti Region) for selected vegetables in a typical growing season. For tomato the usual spraying interval is 7 days, but this is increased to 14 days (Table 2.6) during the long dry season when pest attacks are relatively low.

Table 2.5. Distribution of patterns of pesticide use and management in a sampling ($N = 137$) of vegetable farmers

Variable	Total Respondents	
	Number	Percent
A Timing of Insecticide Application[a]		
Presence of pests	99	39.9
Degree of pest infestation	55	22.2
Date of transplanting	67	27.0
Others	27	10.9
Total	248	100
B Insecticide Handling Practices and Kinds of Protective Cover		
Direction of spraying:		
With the wind	90	65.7
Against the wind	12	8.8
Perpendicular	3	2.2
Do not consider wind direction	32	23.3
Total	137	100
Kinds of protective cover:		
No or partial protective covering[b]	100	74.1
Full protective covering[c]	35	25.9
Total	135	100
C Educational Level of Farmers		
No official education	17	12.7
Primary	14	10.4
Middle	73	54.5
Agric. School/Secondary	17	12.7
Tertiary	13	9.7
Total	134	100
D Farmers' Knowledge of Insecticide Application Rates[a]		
Pesticide label	74	27.5
Agricultural Extension Officer	93	34.6
News (Radio, TV, Newspaper)	21	7.8
Fellow farmer	37	13.7
Pesticide dealer	32	11.9
Others	12	4.5
Total	269	100
E Farmer Re-entry Periods		
Less than 48 h	99	73.3
From 48 to 72 h	19	14.1
More than 72 h	17	12.6
Total	135	100
F Insecticide Storage and Disposal		
Insecticide storage after procurement:		
Safe storage practices	90	67.7
Unsafe storage practices	43	32.3
Total	133	100

Table 2.5. (Cont'd)

Variable	Total Respondents	
	Number	Percent
Disposal of empty insecticide bottles:	0	0
Sell	101	80.2
Throw away on farm	2	1.6
Throw away in town or village	1	0.8
Pile and sell	22	17.5
Bury in ground in farm	0	0
Burn on farm	126	100
Total		
G Sprayer Use and Maintenance		
Type of sprayer used:	114	83.2
Hand pump (Knapsack sprayer)	18	13.1
Motorized sprayer	5	3.6
By Hand	137	100
Total		
Own a sprayer?	58	42.6
Yes	78	57.4
No	136	100
Total		
Wash sprayer after spraying?	125	94.0
Yes	8	6.0
No	133	100
Total		
Disposal of wash water used:	6	4.7
In irrigation canal	113	88.3
On field	5	3.9
In irrigation canal and on field	0	0
In nearby stream	4	3.1
Others	128	100
Total		
H Insecticide Poisoning Cases among Farmers[a]	76	31.0
Headache, dizziness	4	1.6
Vomiting	1	0.4
Unconsciousness	12	4.9
Stomach pain	90	36.7
Weakness	19	7.8
Itching	0	0
Others	43	17.6
None	245	100
Total		

[a] Multiple responses: total responses per item over total respondents; [b] Short trousers/short sleeves or tee-shirt; short trousers/long sleeves; short sleeves or tee-shirt/long trousers; long trousers/long sleeves;   [c] Long trousers, long sleeves, mask and gloves.

Within the vegetable farmer cohort there was an association between age and possible poisoning cases. The apparent association is not reasonably attributable to chance (χ^2 = 13.5, N = 127, DF (degrees of freedom) = 6, P < 0.05). In Figure 2.2 we compare the distribution of possible pesticide poisoning cases between the young (<45 years) and the aged (>45 years). The percentage of farmers reporting body weakness and itching/irritation increased from the young group (39.1 and 6.3% respectively) to the aged group (41.3 and 12.7% respectively). A corresponding downward shift occurred in the percentage of farmers reporting headache/dizziness (i.e. a decrease from 34.4% for the young to 30.2% for the aged). Overall, possible poisoning cases were reported more among the young than the aged farmers. For instance, while about 6% of young farmers said they had not had any possible symptoms of pesticide poisoning, about 14% of the aged group said they had not.

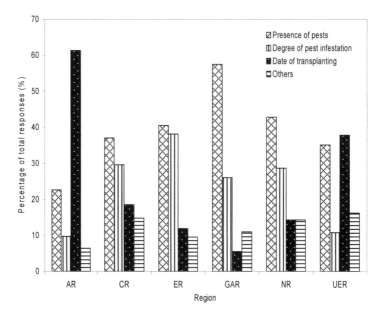

Figure 2.1. Distribution of factors that vegetable farmers consider in timing of insecticide application by region (AR, Ashanti Region; CR, Central Region; ER, Eastern Region; GAR, Greater Accra Region; NR, Northern Region; UER, Upper East Region).

Table 2.6. Spraying records of representative farmers at Akumadan

Case	Cropping months	Interval from planting to first spray (days)	Mean of spraying intervals (days)	Interval from last spray to harvest (days)	Number of sprays
Tomato					
1	Mar-May	7	7	7	12
2	Jul-Sep	21	7	34	6
3	Sep-Nov	14	7	13	10
4	Dec-Feb	14	7	14	9
5	Nov-Jan	14	14	11	6
Pepper (perennial crop)					
1	March	28	21	41	2
Egg-plant					
1	Sep-Dec	28	7	10	12
2	Apr-Jul	28	7	7	12
Okra					
1	Jul-Sep	14	7	40	6
2	Mar-May	14	7	13	10

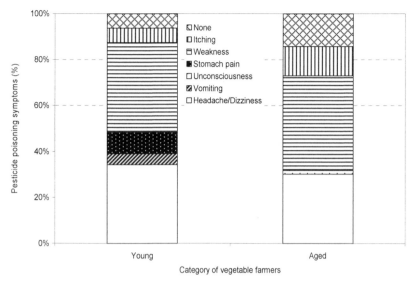

Figure 2.2.   Comparison of distribution of possible pesticide poisoning symptoms for the young
and the aged.

When we assessed for significant interaction between method of pesticide
application and farm size, we found a significant interaction (χ^2 = 46.6, N = 137, DF
= 6, P < 0.001). The distributions of the variables are presented in Figure 2.3.

The majority of the farmers, labourers and their spouses ranked cypermethrin as
the most effective pesticide against insects, followed by lambda-cyhalothrin and
endosulfan in that order (Table 2.7). The average efficacy level for cypermethrin
and lambda-cyhalothrin lay between levels 1 and 2, i.e. providing more than 50
percent insect control. However, some respondents perceived an insecticide as
ineffective or as causing more pest problems. Regarding hazard ranking, about 90%
of the respondents who knew lambda-cyhalothrin, 87% of those who knew
cypermethrin and 84% of those who knew endosulfan ranked them level 1. No
respondent considered lambda-cyhalothrin as harmless (Table 2.8). The average
hazard level for all three chemicals lay between levels 1 and 2, i.e. they were all
considered extremely hazardous.

The hypothesis that a respondent's perception of pesticide hazard is related to its
perceived effectiveness against pests - that is, if it is strong enough to control pests,
then it is hazardous - was tested. The results of the comparison between
respondents' perceptions of pesticide effectiveness and hazard using the χ^2 test do
not support this hypothesis (χ^2 = 6.0, DF = 4, P < 0.05).

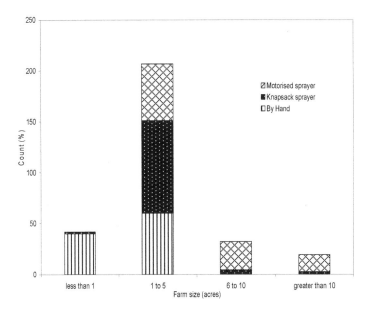

Figure 2.3. Distribution of farm sizes within method of pesticide application

Table 2.7. Vegetable farmers' perceptions of efficacy levels of three commonly known insecticides

Chemical name	Efficacy level[a]					Total responses	Average level
	1	2	3	4	5		
Endosulfan	18	13	7	2	9	49	2.4
Lambda-cyhalothrin	35	11	3	1	2	52	1.5
Cypermethrin	42	5	0	1	6	54	1.6

[a] Levels: 1 = 75-100%; 2 = 50-75%; 3 = below 50%; 4 = not effective; 5 = causes more pest problems

Table 2.8. Vegetable farmers' perceptions of hazard levels of three commonly known insecticides

Chemical name	Hazard level[a]					Total responses	Average level
	1	2	3	4	5		
Endosulfan	63	7	1	1	3	75	1.3
Lambda-cyhalothrin	45	4	1	0	0	50	1.1
Cypermethrin	27	1	0	1	2	31	1.4

[a] Levels: 1 = extremely hazardous; 2 = moderately hazardous; 3 = slightly hazardous; 4 = least hazardous; 5 = no effects.

Discussion

Pests and diseases pose big problems in vegetable production. The damage caused by them has led to farmers using pesticides. Dinham (2003) estimates that 87% of farmers in Ghana use chemical pesticides to control pests and diseases on vegetables. All respondents in the present survey sprayed their crops with pesticides to control pests and diseases. In fact, pesticides are used extensively on vegetable farms, small or large, and farmers use a wide range of chemicals as herbicides, fungicides and/or insecticides. The proportion by class of the pesticides used by farmers in our survey is shown in Figure 2.4. Contrary to many situations in Africa in general, where pesticide usage has been low (as it has always been suggested that the farmer can weed manually), herbicides are the predominant pesticide type in use in vegetable production in Ghana. The preponderance of herbicides over other types of pesticide products could be due to the farmers' perception of weed control (this needs to be investigated further). In our survey the main methods of weed control were identified as manual weeding with hoe and cutlass, and use of herbicides. However, the farmers perceived that herbicides use is able to suppress weeds for a longer time and over a wider area than manual weeding with hoe and cutlass. These together can reduce weeding time and labour cost, especially where there are labour constraints. With the trend towards increasing use of pesticides, farmers will use herbicides even if they can weed manually. Generally, in developing countries, which tend to rely more heavily on herbicides, they do so because of export market (Racke *et al.*, 1997), weed competition and/or labour constraints (Dinham, 2003). Thomas (2003), on the other hand, has mentioned consumer taste (unblemished, cosmetically perfect produce with extended shelf and storage life) and produce yield as factors which increase the use of fungicides and insecticides. Although in Ghana about 13 fungal pathogens are prominent on tomato alone, the use of fungicides was not necessarily to control diseases. Fungicides were used to produce luxuriant vegetative growth. For instance, in the opinion of some farmers the use of mancozeb was to hasten ripening of crops. Those who applied pesticides to control diseases did so, on a trial-and-error basis, because they were not able to identify the pests causing damage.

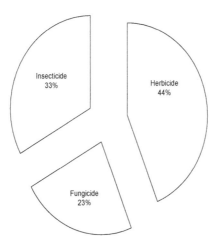

Figure 2.4. Proportion by class of various pesticides used in vegetable production in Ghana

Vegetable farmers sprayed the same very wide range of pesticides on all their crops. For instance, farmers in our survey reported spraying insecticides such as lambda cyhalothrin (Karate 2.5 EC/ULV), cypermethrin (Cyperdin, Cymbush), dimethoate (Perferkthion 400 EC) and endosulfan (Thionex 35 EC/ULV, Thiodan 50 EC) on tomato, pepper, okra, egg-plant (garden eggs), cabbage and lettuce. This information has implications for public health. For instance, endosulfan has a restricted use in Ghana that does not include vegetables (it has only been registered for use on cotton; Table 2.4), yet it is used on vegetables. Although there are no obvious indications of a public health problem as a result of the misuse of endosulfan, the risk is clearly there.

The use of endosulfan on vegetables in Ghana is worrying given its toxicity and persistence in the environment (although more persistent than organophosphates and pyrethroids, endosulfan is not overly persistent). Our study results are corroborated by Ntow (2001), who has demonstrated the presence of organochlorine pesticides and their active metabolites in milk and food crops in Ghana. This is mainly because organochlorines, relative to other classes of pesticides, are resistant to environmental degradation, which allows them to accumulate in plant and animal tissue and to become concentrated in the upper part of the food pyramid (Mbakaya et al., 1994). These observations point to the greater need to monitor foodstuffs for organochlorines and other pesticides in areas where their usage on food crops has been demonstrated (this is the focus of chapter 7 of this thesis).

Pesticides are readily available in agricultural retail stores, and those responsible for retailing them (pesticide dealers) have adopted the extensive practice of measuring out quantities of the pesticides from a larger container. Most of farmers come from poorer sections of the community and, as pesticides are expensive for them, can purchase only relatively small quantities. Farmers also purchase less expensive products, even if they are less suited to the pest(s) requiring control. Information on pesticide application rates comes mainly from Agricultural Extension Officers and/or pesticide labels. To a more limited extent, information also comes from other farmers, pesticide dealers or advertisements (radio, TV, newspaper). To measure pesticides, some farmers used spoons, measuring cylinders, cans and bottles. Others used 'the dose or rate' (i.e. the amount contained in a packet, can or bottle for a given volume of water or for a given acreage, respectively). Wide ranges of rates (both reduced and excessive) were applied on some crops. For instance, in applications on tomato and pepper the rate of lambda cyhalothrin (Karate 2.5 CE) was 4.13 g active ingredient ai/ha, while for chlorpyrifos (Dursban 4E) it varied from 20 to 40 g ai/ha. The recommended rate for lambda cyhalothrin (Karate 2.5 CE) is 12.5 g ai/ha. For chlorpyrifos (Dursban 4E) the recommended rate is 24 g ai/ha. Although it is undeniable that vegetable crops need large quantities of pesticides for control of their pests and diseases, it remains doubtful whether all the sprays are really necessary. A high level of pesticide use is unnecessary as well as harmful. Integrated pest management (IPM) and organic agricultural strategies produce comparable yields, but adoption of these approaches remains low so far for a variety of reasons (Danso et al., 2002).

Chemical pesticides were usually sprayed in combination, and the efficacy of one may mask the inefficacy of others in the mixture. Mixing of pesticides was encouraged by the farmers' desire to have rapid knockdown of pests. This idea is questionable (Medina, 1987), at least as practised, because the combinations used are indiscriminate. The practice of using indiscriminate combinations of pesticides, particularly of insecticides, may have contributed to the increase in incidences of insect pest infestation of tomato in Ghana (Biney, 2001). The practice defies some of the basic principles of insecticide management. For instance, Metcalf (1980), in his recommendation of strategies for pesticide management, states that the use of mixtures of insecticides must be avoided, since mixtures of insecticides generally result in the simultaneous development of resistance.

The number of sprays per crop season, however, varied widely among crops, locations and the farmers interviewed in the survey. For instance, in a tomato-growing season of about 90 days at Akumadan in Ashanti Region, farmers sprayed an average of 6-12 times with three to six pesticides on a calendar basis. The usual spraying interval was 7 days. The spraying regime of the farmers also varied between climatic seasons. Farmers utilised more sprays during the wet season when pests and diseases proliferated. Besides, increased wash-off by rainfall necessitated further applications of pesticides. In Uganda (Tukahirwa, 1987), Mubuka farmers sprayed phosphamidon on tomato at a frequency of up to once per week to maturity. Vegetable growers surely need good advice on which to base their pest and disease control programmes.

The investigations showed that, farmers' opinions on the direction of spraying varied. While some farmers considered the wind direction during spraying and therefore sprayed with the wind direction, others did not. The latter category of farmers sprayed along rows, back and forth, even against the wind or perpendicular to the wind direction. Where this occurs, the wind may blow the chemicals into the body, including the face, of the farmer. This poor spraying practice presents great potential for exposure of farmers to chemicals from both skin contact and inhalation.

Farmers used very little personal protection during spraying (Table 2.5). Some (25.9%), especially those within cooperatives, did use some protective clothing. This included rubber boots, a coverall with long sleeves, gloves and a piece of cloth to cover the mouth. The majority wore trousers and a long-sleeved shirt. However, some wore a short-sleeved shirt and short trousers, with no gloves, and barefooted farmers (they wore slippers which exposed a greater part of their feet) even used their bare hands to mix pesticides in a container. As a consequence, their legs, feet and hands came into contact with pesticides. About 80% of the farmers surveyed had become ill from pesticide exposure. The most frequent symptoms were reported as weakness, headache and/or dizziness.

With regard to pesticide application procedures, the knapsack was the most popular spraying equipment used, though a few farmers did use motorised sprayers. Lack of capital was the main reason why farmers tended to use the knapsack sprayer. In some cases (3.6%), farmers used a brush, broom or leaves tied together to splash pesticides from a bucket. This was not surprising, since over 50% of respondents did not own a sprayer. Where this practice occurs, users are clearly exposed to the pesticide, especially as very few can afford or have available protective clothing. The use of the knapsack sprayer in itself presents some danger to the user, since it is prone to leakage, especially as the sprayer ages. Matthews *et al.* (2003) have identified causes of leakage from the knapsack and have emphasised the need to provide better-quality equipment at an acceptable cost that will be more durable in a hot and humid tropical environment such as Africa. In addition, during spraying, farmers do not distinguish between target and non-target crops. Non-target susceptible crops are therefore also exposed to the pesticide.

The commonest way of disposing of sprayer wash water (88.3%) and empty pesticide containers (80.2%) among the farmers interviewed was by throwing them on the field. As one walks at the edges of farms, one can clearly see pesticide containers lying about (over 60% of farms). Where farms are close to waterways (which is the case in many farming communities), the disposal of unwanted pesticide solutions and empty containers in the field presents a potential pollution problem for aquatic systems, which are sources of livelihood for human communities and support varied animal and plant life. Non-target flora and fauna may concentrate these chemicals in their tissues and pass them along the food chain. The accumulation of such pollutants in the food chain may restrict the consumption of valuable food resources such as fish. There is therefore a need for carefully conducted scientific studies to assess the degree of environmental contamination due to pesticides (this is the focus of chapter 4 of this thesis).

The results of the ranking game showed that farmers could differentiate pesticides in terms of their hazard to human health. Both experts and farmers believe, when explicitly asked, that pesticides are hazardous. However, farmers' perception of relative hazard differs slightly from that of experts using the WHO ranking. For instance, while farmers classify endosulfan, lambda-cyhalothrin and cypermethrin as extremely hazardous, WHO classifies all of them as moderately hazardous. Inhalation and irritation were perceived as the main indicators of a pesticide's hazard level. Farmers would therefore be expected to be more concerned with avoiding pesticide inhalation or skin exposure. Unfortunately, fewer than 30% of the respondents wore full protective covering, which shows the difference between passive and active (decision-making) knowledge.

Acknowledgements

The authors are grateful to the Dutch Government (through the UNESCO-IHE Institute for Water Education), the International Water Management Institute (IWMI) and the International Foundation for Science (IFS) for their financial support of this research. Vegetable farmers in Ghana are acknowledged for their cooperation in making the study successful.

References

Biney PM, Pesticide use pattern and insecticide residue levels in Tomato (*Lycopersicum esculentum*) in some selected production systems in Ghana. *Mphil Thesis*, University of Ghana, Legon, Ghana, 127 pp (2001).

Clarke E, Levy LS, Spurgeon A and Calvert LA, The problems associated with pesticide use by irrigation workers in Ghana. *Occupational Medicine* **47(5)**: 301-308 (1997).

Danso G, Fialor SC and Drechsel P, Perceptions of organic agriculture by urban vegetable farmers and consumers in Ghana. *Urban Agricultural Magazine* **6**, 23-24 (2002).

Dinham B, Growing vegetables in developing countries for local urban populations and export markets: problems confronting small-scale producers. Pest Manag Sci. 59: 575-582 (2003).

Matthews G, Wiles T and Baleguel P, A survey of pesticide application in Cameroon. *Crop Protection* **22**; 707-714 (2003).

Mbakaya CFL, Ohayo-Mitoko GJA, Ngowi VAF, Mbabazi R, James M, Simwa JM, Maeda DN, Stephens J and Hakuza H, The status of pesticide usage in East Africa. *African Journal of Health Sciences* **1 (1)**: 37-41 (1994).

Medina, C.P., Pest Control Practices and pesticide perceptions of Vegetable farmers in Loo Valley, Benguet, Philippines. *In: Management of Pests and Pesticides: Farmers' Perceptions and Practices,* ed by Tait J and Napompeth B, Westview Press, London, pp 150-157 (1987).

Mensah E, Amoah P, Abaidoo RC and Drechsel P, Environmental concerns of (peri-) urban vegetable production – Case studies from Kumasi and Accra. *In*: Waste Composting for Urban and Peri-urban Agriculture – Closing the rural-urban nutrient cycle in sub-Saharan Africa, ed by Drechsel P and Kunze D, IWMI/FAO/CABI: Wallingford, 55-68 (2002).

Metcalf RL, Changing role of insecticides in crop protection. *Ann. Rev. Ent.* 25(22): 119-256 (1980).

Ntow WJ, Pesticide misuse at Akumadan to be tackled. *NARP Newsletter,* 3(3): 3 (1998).

Ntow WJ, Organochlorine Pesticides in Water, Sediment, Crops and Human fluids in a farming community in Ghana. Arch. Environ. Contam. Toxicol., 40(4): 557-563 (2001).

Racke KD, Skidmore MW, Hamilton DJ, Unsworth JB, Miyamoto J and Cohen SZ, Pesticide fate in tropical soils. Pure & Appl. Chem., 69(6): 1349-1371 (1997).

Thomas MR, Pesticide usage in some vegetable crops in Great Britain: real on-farm applications. *Pest Manag Sci* 59: 591-596 (2003).

Tukahirwa EM, Pest Control Practices and pesticide perceptions of vegetable farmers in Loo Valley, Benguet, Philippines. *In: Management of Pests and Pesticides: Farmers' Perceptions and Practices*, ed by Tait J and Napompeth B, Westview Press, London, pp 182-190 (1987).

Warburton H., Palis FG and Pingali PL, Farmer perceptions, Knowledge and pesticide use practices. In: Impact of pesticides on farmer health and the rice environment, ed by Pingali PL and Roger PA, Kluwer Academic Publishers, Massachusetts, 59-95 (1995).

Chapter 3

Dissipation of Endosulfan in Field-Grown Tomato (*Lycopersicon esculentum*) and Cropped Soil at Akumadan

Submission to a scientific journal based on this chapter:

William J. Ntow, Jonathan Ameyibor, Peter Kelderman, Pay Drechsel and Huub J. Gijzen. Dissipation of Endosulfan in Field-Grown Tomato (*Lycopersicon esculentum*) and Cropped Soil at Akumadan, Ghana. *J. Agric. Food Chem.* Accepted for publication on October 23, 2007.

Dissipation of Endosulfan in Field-Grown Tomato (*Lycopersicon esculentum*) and Cropped Soil at Akumadan

Abstract

The dissipation and persistence of endosulfan (6,7,8,9,10,10-hexachloro-1,5,5a,6,9,9a-hexahydro-6,9-methano-2,4,3-benzodioxathiepin 3-oxide) applied to field-grown tomato (*Lycopersicon esculentum*) were studied at a vegetable-growing location in Ghana. Plant tissue samples and cropped soil collected at 2 h-14 days and 8 h-112 days, respectively, after application, were analyzed by gas chromatography-electron capture detection (^{63}Ni) to determine the content and dissipation rate of endosulfan isomers (α- and β-endosulfan) and the major metabolite, endosulfan sulfate. After two foliar applications of commercial endosulfan at 500 g of active ingredient/hectare, the first-order reaction kinetic was confirmed to describe the dissipation of endosulfan residues in tomato foliage and cropped soil. However, functions that best fit the experimental data were the biphasic process for foliage and the monophasic process for cropped soil. Calculated DT_{50} and DT_{90} values for endosulfan residues in cropped soil were not significantly ($p < 0.05$) different for each of the two isomers.

Keywords: Dissipation; endosulfan; half-life; persistence; soil; tomato (*Lycopersicon esculentum*)

Introduction

Chlorinated organic pesticides are one of the major groups of chemicals responsible for environmental contamination. Many chlorinated pesticides are highly toxic and considered a potential risk to both human health and the environment. Technical endosulfan, a mixture of two stereoisomers, that is, α- and β-endosulfan (Figure 3.1a,b) in the approximate ratio of 7:3 (Shetty *et al.*, 2000; Kennedy *et al.*, 2001), is a chlorinated pesticide for control of a large spectrum of insect pests on a wide range of crops (Aguilera-del Real *et al.*, 1997). It is used in many countries throughout the world for the control of pests on fruits, vegetables, tea, tobacco, and cotton (Antonious and Byers, 1997; Sethunathan *et al.*, 2002). Because of such abundant usage, and the potential for accumulation in the environment (endosulfan is not readily detoxified by soil microorganisms), residues are detectable in soils, sediments, and crops at harvest time (Goebel *et al.*, 1982; U.S. Department of Health and Human Services, 1990). Although the metabolites of endosulfan, that is., sulfate, diol, ether, hydroxy ether, and lactone, have been shown to occur (Maier-Bode, 1968; Schuphan *et al.*, 1968), only the sulfate metabolite (Figure 3.1c) is significant as a residue (Antonious and Byers, 1997).

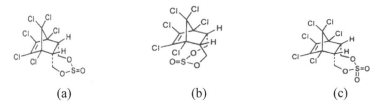

$$(a) \qquad\qquad (b) \qquad\qquad (c)$$

Figure 3.1. Endosulfan isomers and its sulfate metabolite: (a) α-endosulfan; (b) β-endosulfan; (c) endosulfan sulfate

Endosulfan, which has been found in residue monitoring (Osafo and Frimpong, 1998; Ntow, 2001, 2005) and food crops studies (Amoah *et al.*, 2006), is one of the commonly used chlorinated pesticides on vegetables in Ghana (Ntow *et al.*, 2006). At Akumadan (1° 57' W, 7° 24' N), a prominent vegetable-farming community in Ghana (Ntow, 2001), the pesticide is one of the predominant active ingredients used for controlling leaf miners, bollworm, fruit fly, *etc.*, on tomato and has the potential of environmental contamination because of much overuse, abuse, and misuse of the pesticide (Ntow *et al.*, 2006). One aspect of the range of studies needed to assess the environmental impact of a pesticide is environmental fate studies, and understanding the persistence and dissipation of the pesticide is an important step forward.

This research uses field experiments to provide insight into the persistence and dissipation of endosulfan applied to field-grown tomato in sandy loam soil under the tropical conditions of Ghana and to answer the following questions: (1) how is endosulfan distributed, qualitatively and quantitatively, in tomato field following spraying on plant foliage using 1.5 times the labelled rate? (2) Does endosulfan persist in tomato foliage and fruit as total residues to which consumers and/or insects may be exposed? Finally, what are the persistence and the dissipation rate of endosulfan isomers and major metabolite in tomato foliage and cropped soil?

Materials and Methods

Reagents
Analytical standards of α-endosulfan (96.0% purity), β-endosulfan (98.0% purity), and endosulfan sulfate (97.5% purity) were supplied by Dr. Ehrenstorfer GmbH, Augsburg, Germany. Thionex 35 EC/ULV containing 350 g/L ai of endosulfan (α:β = 1.96:1) was obtained from Hoechst Agrevo Ltd. through a local pesticide dealer at Akumadan. Stock solutions of α- and β-endosulfan and endosulfan sulfate (100 µg/mL) were prepared separately in *n*-hexane. All organic solvents used were of GC grade (Sigma, München, Germany; or BDH, VWR International, Poole, U.K.).

Experimental design and endosulfan treatment
The field study was conducted during February–May 2005 at Akumadan. Prespray soil samples (about 200 g dry weight) from 0-10, 10-20, 20-30, and 30-40 cm depths of the experimental field were collected with a corer (5.0 cm diameter) at random and analysed for water content, pH, organic matter content, texture (clay, silt, and sand contents), and bulk density. Conventional soil analyses were carried out using standard methods. Water content was determined by weight loss after drying in an oven at 110 °C. Measurement of pH (1:2.5 soil/water) was made using a Hach model pH-meter. Soil organic matter was determined according to the Walkley-Black method (Jackson, 1958); particle size distribution was determined by using the pipet method (Day, 1965) [three sizes were estimated: <0.002 mm (clay), 0.002-0.02 mm (silt), and >0.02 mm (sand)]; and bulk density was determined by weight loss over volume of a cylinder after drying in an oven at 102 °C.

The soil was a sandy loam having 68.5% sand (>0.02 mm), 21.6% silt (0.002–0.02 mm), and 9.9% clay (<0.002 mm); pH 6.5; organic matter of 1.3%; water content of 9.5%; and bulk density of 1.52 g/cm^3 for all segments. Of about 4000 m^2 area prepared for field studies (no endosulfan had been sprayed on the field for over 2 years since September 2002), nine plots each measuring 15 x 15 m were demarcated in a 3 x 3 randomised complete block design for two treatments (T1 and T2) and a control treatment (TC), leaving a border area of about 2.5 m around the plots. Each treatment was replicated three times. On February 9, 2005, 14-day-old tomato seedlings were transplanted at 75 cm apart in rows (row to row distance of 60 cm; 450 plants per plot). On March 12, 2005 (i.e., 31 days after transplanting), endosulfan (Thionex 35 EC/ULV; labeled rate is 2.1 L/ha) was applied on T1 plots from a height of 20-25 cm above the plant canopy at a rate of 3 L/ha (500 g of ai/ha) in 215 L of water on tomato (*Lycopersicon esculentum* var. Power), using a portable backpack sprayer [Knapsack CP 15 L equipped with one conical nozzle operated at 40 psi (275 kPa)]. Before use, the spraying device was calibrated with respect to homogeneity of the spray beam and pumping volume per time unit. The application of pesticide to the plots was executed bandwise and in a criss-cross pattern to ensure a uniform distribution. On March 12 and April 9, 2005 (i.e., 31 and 59 days after transplanting, respectively), the above endosulfan treatment was applied on T2 plots. The second spraying was done at the time of about 50% fruit formation. TC plots were kept as the untreated controls.

Throughout the experiment, the plots were kept free of weeds by hand hoeing, taking care not to disturb the upper layer of soil. The plots were irrigated four times (furrow irrigation). Water was pumped from a reservoir into a head ditch from where the water flowed by gravitation into furrows running across the field. The irrigation water was crosschecked for endosulfan residues. The first irrigation was given on the day of transplanting and thereafter at 21 day intervals. During the experimental period, there was a single rainfall event on April 11, 2005, estimated at 11 mm. Mean relative humididty was 71%; maximum and minimum temperatures averaged 30 and 25 °C, respectively, with a mean of 27 °C during the experimental period.

Sampling for endosulfan residues analysis
After treatment T1, samples of leaves (10 g of fresh weight each) were drawn from each replicate plot at intervals of 0 (2 h after spray), 1, 2, 6, and 14 days. Triplicate tomato leaf samples were collected randomly from the midcanopy of plants from each replicate plot. In addition, whole fruit, root, and stem samples (100-200 g of fresh weight each), and leaf (10 g of fresh weight) samples were taken randomly at same time intervals from treatment T2 plots.

Soil samples were drawn at time intervals of 0 (8 h after spray), 1, 2, 6, 14, 28, 56, and 112 days after application from four different levels (0-10, 10-20, 20-30, and 30-40 cm depths) from treatment T1 plots and additionally from the surface (0-10 cm) of treatment T2 plots at intervals of 0 (2 h after spray), 1, 2, 6, and 14 days. One-metre-sided squares were delimited randomly in the 15 x 15 m subplot. In each square, soil cores were taken from the specified soil depths. Thus, within a 15 x 15 m subplot, variable numbers of soil samples (3-10) were taken at the specified times above. All core samples of different depths and treatments were homogenized separately. Each soil sample was a composite of 36 cores (5.0 cm diameter) from 0-10, 10-20, 20-30, and 30-40 cm depths (T1 plots) and 0-10 cm depth (T2 plots). Three replicate samples, about 50 g each, were drawn from each composite. Sorted samples of soil, leaf, fruit, stem, and root were wrapped in aluminum foil, packed in polythene bags, and transported to the CSIR Water Research Institute Laboratory in Accra (a distance of 349 km) within 24-48 h on ice in clean ice chests. Upon arrival at the laboratory, leaf, fruit, stem, and root samples (peel and flesh) were given a cold water wash with a soft brush to remove adhering soil particles and subsequently kept in a freezer at –4 °C until required for extraction, which was carried out within 24 h. Sorted soil samples were transferred into pans to air-dry at ambient temperature.

Analytical procedures

Samples of tomato plant parts (roots, stems, leaves, and fruits) were extracted according to procedures described in FAO/IAEA (1997). Briefly, the frozen samples were thawed, and each plant part (approximately 5 g of fresh weight) was cut (or chopped in the case of leaf) into small pieces and homogenised in a mortar. The plant parts were transferred to a pre-extracted Whatman cellulose extraction thimble; lipids were extracted for 8 h with methanol (200 mL) in a Soxhlet apparatus cycling four or five times per hour. The extract was passed through a preconditioned SPE column (Bond Elute C-18 3-cc/500 mg; Varian, Palo Alto, CA, USA) as described in Ntow (2001). Residues trapped in the column were eluted with *n*-hexane (1.5 mL) into a glass vial and brought to volume (2 mL) with *n*-hexane for analysis by gas chromatography.

Air-dried soil samples were ground in a mortar and sieved (2 mm). About 5 g representative sieved samples were weighed into extraction thimbles and extracted for 8 h with methanol (200 mL) in a Soxhlet apparatus as described above for plant samples. The extract was passed through a preconditioned SPE column and treated in the same way as described above for plant parts.

Analyses were performed with a Perkin-Elmer AutoSystem gas chromatograph equipped with a ^{63}Ni electron capture detector. Separations were on a 30 m x 0.32 mm i.d. capillary column with 0.25 µm methyl phenyl phase (Perkin-Elmer Elite-225). The gas flow (helium) was set to 16 mL/min through the column and at 30 mL/min makeup (nitrogen) through the detector. Sample volumes (1 µL) were injected in a split mode at 250 °C, and the oven temperature was programmed as follows: 100 °C for 1 min, increased to 150 °C (10 °C/min), 250 °C (5 °C/min), then at 30 °C/min to 300 °C (held 10 min). The detector temperature was 350 °C. The retention times (RT) of each of the endosulfan isomers and the sulfate metabolite were compared with those of the external standards, and the data were recorded. The RT of α- and β-endosulfan and endosulfan sulfate were observed as 22.6, 25.2, and 26.8 min, respectively.

In recovery experiments, soil and plant samples from the control plot fortified at levels of 0.1 and 0.5 mg/kg were used. Each fortification level was prepared in three replicates. Chopped or cut plant parts and sieved soil samples were placed in 250 mL standard joint borosilicate bottles and fortified by the addition of appropriate volumes of previously prepared stock solutions of endosulfan. The bottles were capped, manually shaken to ensure thorough mixing, and stored in a deep freezer at –4 °C for 24 h to simulate residue sample storage conditions. The recovery values (mean ± SE) observed were 102.5 ± 2.1, 87.6 ± 1.3, and 84.5 ± 1.2% for α- and β-endosulfan and endosulfan sulfate, respectively, using a fortified soil sample, whereas these values were 85.6 ± 1.3, 84.3 ± 0.9, and 86.5 ± 1.1% for plant samples. Residue data were not corrected for efficiency of recovery. The limit of quantification was 0.001 mg/kg for each of the endosulfan isomers and the metabolite sulfate.

Residues concentrations in soil (milligrams per kilogram of dry weight) and plant samples (milligrams per kilogram of fresh weight) (treatment T2) were converted to loadings per field area in milligrams per hectare. To determine the total mass of active ingredient in soil and plant the following calculations were used:

$$M_{as} = C \times \text{assumed area sprayed} \qquad\qquad (3.1)$$

$$M_{ap} = C \times \text{assumed area sprayed} \times \text{crop yield} \qquad\qquad (3.2)$$

M_{as} = total mass of active ingredient in soil; M_{ap} = total mass of active ingredient in plant; C = concentration of unit sample in each constituent (mg/kg/m^2 of soil or mg/kg/ plant); assumed area sprayed = (15 × 15 m plot size) × 3 number plots; and crop yield = 450 plants per plot (15 × 15m area). To determine the mass of active ingredient per hectare, it is multiplied by 14.8. From the residue loadings in the samples it was possible to estimate the proportion of endosulfan in each constituent of the tomato field ecosystem in one season.

The dissipation of endosulfan in tomato foliage and cropped soil was determined by a nonlinear regression of the pesticide residue concentration against time (treatment T1) implemented in Microsoft Excel. The statistical parameters, r^2, k, and C_o were determined using an iterative nonlinear regression procedure using SPSS software (SPSS software, version 12.0.1 for Windows, SPSS Inc., Chicago, IL). DT$_{50}$ and DT$_{90}$ values for α- and β-isomers were also calculated.

Results

In Table 3.1, endosulfan residue contents in and their distribution among leaves, fruits, stems, roots, and soil in the course of the experiment (T2 plots) are presented. Shortly after treatment, highest total endosulfan residue contents were found in the leaves, followed by fruits, soil, stems, and roots. Among the plant parts, leaves had the highest content of total endosulfan residues, followed by fruits, stems, and roots. For tomato leaves a sharp decline in the total endosulfan contents was observed within 24 h, followed by a relatively slow decline to the termination of the experiment.

Table 3.1. Levels and Distribution of Endosulfan (α-Endosulfan, β-Endosulfan, Endosulfan Sulfate, and Sum) in Tomato Plant Parts (Treatment T2)

sample	residue (mg/kg) [a]			
	time (days) [b]	α	β	sulfate	Sum [c]
control	0	<0.001 [d]	<0.001	<0.001	<0.001
leaves	0	1.110	0.702	0.039	1.851
	1	0.122	0.362	0.086	0.570
	2	0.050	0.085	0.103	0.238
	6	0.017	0.016	0.116	0.149
	14	0.008	0.006	0.020	0.034
fruits	0	0.446	0.428	0.030	0.904
	1	0.297	0.291	0.052	0.647
	2	0.210	0.173	0.266	0.642
	6	0.062	0.037	0.091	0.190
	14	0.031	0.022	0.050	0.103
stems	0	0.126	0.124	0.010	0.260
	1	0.059	0.056	0.055	0.170
	2	0.031	0.026	0.073	0.130
	6	0.006	0.004	0.071	0.081
	14	0.003	<0.001	0.027	0.030
roots	0	0.064	0.047	<0.001	0.111
	1	0.027	0.043	<0.001	0.070
	2	0.015	0.035	<0.001	0.050
	6	<0.001	0.010	0.020	0.030
	14	< 0.001	0.003	0.007	0.010
soil	0	0.574	0.221	<0.001	0.795
	1	0.527	0.204	<0.001	0.731
	2	0.527	0.199	<0.001	0.726
	6	0.213	0.067	0.161	0.441
	14	0.057	0.042	0.027	0.126

[a] Mean of triplicate analyses from three replicates. [b] Reference to the second application for T2. [c] α-Endosulfan + β- endosulfan + endosulfan sulfate. [d] Limit of quantification.

Treatment T2

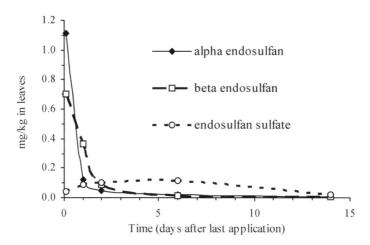

Figure 3.2. Dissipation of α- and β-endosulfan and endosulfan sulfate in leaves. Maximum concentration of endosulfan sulfate was measured on day 6, which was about 6% of the initial amount of total endosulfan.

Treatment T2

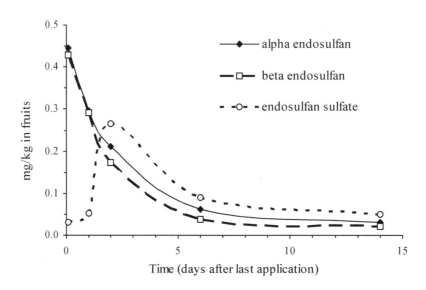

Figure 3.3. Dissipation of α- and β-endosulfan and endosulfan sulfate in fruits. Maximum concentration of endosulfan sulfate was measured on day 2, which was about 30% of the initial amount of total endosulfan.

In tomato fruits, initial total endosulfan residues was above 0.50 mg/kg 2 h after treatment and consisted primarily of the α- and β-isomers, whereas only a relatively trace residue level of endosulfan sulfate metabolite was detected 2 h following treatment. Residues of the sulfate metabolite 2 days after treatment constituted about 30% of the initial amount of total endosulfan in tomato fruits (Figure 3.3).

As evident from the data (T1 plots) given in Table 3.2 endosulfan parent isomers and their sulfate breakdown product did not move beyond a 30 cm depth. Endosulfan α-isomer remained confined in the 0-10 cm layer. The β-isomer of endosulfan leached down to 30 cm until 28 days of experimentation. The amount present in the 20-30 cm layer was markedly lower than that in the 0-10 cm layer. Endosulfan sulfate metabolite did not leach beyond 10 cm.

A chemical balance budget made using the data in Table 3.1 (T2 plots) from a late application of endosulfan (April 9, 2005) (at time $t = 14$ days) showed that most of the residues of the pesticide were found in the tomato plant system (74%), whereas only a relatively small proportion (26%) was found in cropped soil of the amount (0.5%) that remained on-field. In the tomato plant system, the distribution of total endosulfan residues followed the order fruits (43%) > leaves (14%) > stem (13%) > root (4%).

Discussion

In the interpretation of the results of this study, the word 'in' has been used to mean 'in', 'on', or 'in and on'. Our study did not differentiate whether residues were situated on outer surfaces of plant tissues or were taken up into the tissues of the plant. However, several authors have reported endosulfan residues, including metabolites, in, on, or in and on tomato plant tissues (Aguilera-del Real et al., 1997; Antonious et al., 1998; Gonzalez et al., 2003), as well as in, on, or in and on tissues of other plants and in soil (Bead and Ware, 1969; Stewart and Cairns, 1974; Chopra and Mahfouz, 1977; Martens, 1977; Singh et al., 1991; Raha et al., 1993; Kathpal et al., 1997; Kennedy et al., 2001).

Shortly after treatment of endosulfan on field-grown tomato using Thionex (35 EC/ULV) formulation, the levels of total endosulfan residues were markedly higher in leaves than in fruits, stems, or roots due, partly, to the foliar application of the pesticide. The other reasons could be the horizontal position of the lamina of the leaves as well as the difference in surface area between leaves and other tissues of the plant (Raha et al., 1993; Antonious et al., 1998). However, at the termination of the experiment, endosulfan residue levels were about 3-10 times higher in fruits than in other tissues. Miglioranza et al. (1999) found that high carotenoid levels (lipophilic substances) are responsible for retaining chlorinated hydrocarbons in the body and peel of vegetables, and we believe this to explain the higher endosulfan residue levels in fruits than in leaves at the termination of the experiment.

The measurement of endosulfan (a persistent organochlorine compound) in tomato fruit (a vegetable crop) is of great importance as its uptake is a major pathway for a toxic substance into the food chain leading to human exposure (Donker et al., 1994). The Codex Committee on Pesticide Residues (CCPR) considers a total endosulfan concentration of 0.50 mg/kg in tomato to be the maximum residue level (MRL) [CCPR (https://secure.pesticides.gov.uk/MRLs)]. After treatment, the residue level of total endosulfan in fruit was 0.90 mg/kg, and at harvest time, that is, 2 weeks later (according to the preharvest interval), it was 0.10 mg/kg, which is markedly lower than the Codex MRL.   However, a definite conclusion on the safety of the consumption of field-grown tomato cannot be reached because high application rates of endosulfan (1000 g ai/ha, 5-10 kg per seasonal total for tomato by some farmers at Akumadan) are reported (Ntow, unpublished results). The human dietary exposure to pesticides from the consumption of vegetables is the subject of chapter 7.

The endosulfan formulation used (Thionex) contained two isomers, α- and β-isomers, with a higher relative amount of α-isomer than β-isomer (α-isomer/β-isomer = 1.96:1). During the first 2 h following pesticide treatment, the residue level of α-endosulfan was higher than that determined for β-endosulfan in leaves, but from day 2, residue levels found for both isomers were, in general, similar (Figure 3.2; Table 3.1). In fruits, residue levels of α- and β-endosulfan were very similar during the entire 14 day period of experimentation, although levels of α-endosulfan were, in general, higher than those determined for β-endosulfan (Figure 3.3; Table 3.1). According to Kimber et al. (1994), Kathpal et al. (1997) and Antonious et al. (1998), although the endosulfan α-isomer is about 70% of the active ingredient in commercial formulations, it is found in solid surfaces at appreciable levels only immediately after spraying, due to its high volatility. Endosulfan α-isomer is more volatile (vp = 0.006 mmHg at 20 °C) and less water-soluble (2.29 mg/L at 22 °C) compared to the β-isomer (vp = 0.003 mmHg at 20 °C and water solubility = 31.1 mg/L at 22 °C) (Guerin and Kennedy, 1992; Antonious et al., 1998). Endosulfan was converted to the sulfate metabolite in foliage and fruit of tomato in 2 h, following treatment. In both foliage and fruit, this breakdown product of endosulfan persisted until the termination of the experiment at 14 days. Oxidation of the parent compounds (Kennedy et al., 2001) causes an initial buildup in the sulfate metabolite, which reaches a peak in 6 and 2 days in foliage and fruit, respectively, after application. Given that endosulfan sulfate is formed in many natural environments through biological oxidation and that it is only slowly degraded, both chemically and biologically (Miles and Moy, 1978), it may represent a predominant residue of endosulfan in aerobic environments.

To describe the dissipation of residues of endosulfan isomers in tomato foliage and cropped soil, a monophasic dissipation model in first-order kinetics derived from equation 3.3 (Morrica et al., 2002) was used:

$$C_t = C_o e^{-kt} \tag{3.3}$$

C_o is the y-intercept value, C_t the concentration of endosulfan residues in matrix at time t (mg/kg), t is the postapplication time (days), and k is the slope of the dissipation line. DT_{50} and DT_{90} values and the dissipation rate constant (k) were determined from the slope of a nonlinear regression plot of C_t versus t.

Table 3.2. Dissipation of Endosulfan in tomato cropped soil in T_1 (500g of active ingredient given 31 days after transplanting)

days after spraying	depth (cm)	endosulfan residues (mg/kg) [a]				
		α-endosulfan	β-endosulfan	endosulfan sulfate	Σ endosulfan	dissipation (%)
control plot	0-40	<0.001 [b]	<0.001	<0.001	<0.001	-
0 (8 h)	0-10	2.30 ± 0.34	0.88 ± 0.11	<0.001	3.18 ± 0.23	-
1	0-10	2.11 ± 0.11	0.82 ± 0.07	<0.001	2.92 ± 0.18	8.1
2	0-10	2.11 ± 0.06	0.80 ± 0.07	<0.001	2.90 ± 0.12	8.8
6	0-10	0.85 ± 0.11	0.27 ± 0.01	0.65 ± 0.08	1.77 ± 0.19	44.5
14	0-10	0.23 ± 0.01	0.13 ± 0.01	0.11 ± 0.01	0.47 ± 0.01	
	10-20	<0.001	0.03 ± 0.00	<0.001	0.03 ± 0.00	
	20-30	<0.001	0.01 ± 0.00	<0.001	0.01 ± 0.00	
					0.51	84.3
28	0-10	<0.001	0.04 ± 0.00	0.11 ± 0.01	0.15 ± 0.01	
	10-20	<0.001	<0.001	<0.001	<0.001	
					0.15	95.2
56	0-10	<0.001	0.01 ± 0.00	0.07 ± 0.01	0.08 ± 0.01	
	10-20	<0.001	<0.001	<0.001	<0.001	
					0.08	97.6
112	0-10	<0.001	<0.001	0.04 ± 0.00	0.04 ± 0.00	
	10-20	<0.001	<0.001	<0.001	<0.001	
	20-30	<0.001	<0.001	<0.001	<0.001	
	30-40	<0.001	<0.001	<0.001	<0.001	
					0.04	98.7

[a] **Mean (± SD) of three replicates.** [b] **Limit of quantification**

In tomato cropped soil, endosulfan dissipation followed an essentially first-order kinetic. As can be seen in Figure 3.4a in cropped soil the concentration of endosulfan gradually decreased with time during the study period of 120 days. The calculated DT_{50} and DT_{90} values for endosulfan in tomato cropped soil were not significantly ($p < 0.05$) different for each of the two isomers [4.31 (± 0.105), 14.3 (± 0.105) days for the α-isomer, respectively; 4.31 (± 0.0255), 14.3 (± 0.0255) days for the β-isomer, respectively] in cropped soil (Figure 3.4a,b). These findings suggest that, in this experiment, α-endosulfan and β-endosulfan did not differ in persistence in cropped soil.

However, in tomato foliage, endosulfan concentration also decreased with time, but more rapidly initially and then slowly (Figure 3.2). This deviation of foliage dissipation kinetic from first-order kinetic, with exhibition of biexponential (two-stage) dissipation kinetic, has been often observed for endosulfan (Antonious et al., 1998; Kennedy et al., 2001). Some authors (Ghadiri et al., 1994; Kennedy et al., 2001) explain this biphasic model through an initial rapid volatilisation phase followed by a slower rate of dissipation. The high volatilisation rate of endosulfan has been reported from solid surfaces as well as aqueous systems (Beard and Ware, 1969; Singh et al., 1991; Guerin and Kennedy, 1992).

Thus, in tomato foliage, dissipation of endosulfan was described by a biphasic model (Morrica et al., 2002)

$$C_t/C_o = ae^{-k_1 t} + (1 - a)e^{-k_2 t} \tag{3.4}$$

C_o is the initial concentration of endosulfan (mg/kg), C_t is the concentration at time t (mg/kg), t is the post application time (days), k_1 and k_2 are fast and slow dissipation rate constants, and a is a constant (Morrica et al., 2002). Relatively better correlation coefficients were obtained when the dissipation was fitted to two nonlinear phases. Figure 3.5 shows the nonlinear relationships together with the values of the statistical parameters calculated for endosulfan parent isomers and total endosulfan using the model. The biphasic shape of endosulfan dissipation curves had been reported earlier by Kathpal et al. (1997), Antonious et al. (1998), and Kennedy et al. (2001) to describe the two-phase dissipation of pesticides in foliage and soils, when an initial period of fast pesticide loss is followed by a phase of slower dissipation.

To assess the vertical movement of endosulfan, soil core concentrations were measured to judge the pesticide content in different soil layers in relation to the applied amount (Table 3.2). Endosulfan was not detected beyond a 30 cm depth of soil at Akumadan. We attributed this finding to its high soil adsorption coefficient, $K_{oc} = 12400$ [EXTOXNET (http://extoxnet.orst.edu/pips/ghindex.html)], which presumably led to concentrations below the quantification limit in the subsoil layers.

The results of the chemical balance budget (at time $t = 14$ days) after two foliar applications (T2) of endosulfan (total oad of 1000 g of ai/ha) as Thionex (35 EC/ULV) formulation indicates that a greater percentage (99.5%) of endosulfan dissipates from a tomato field with only a small percentage (0.5%) remaining on-field

α-Endosulfan in soil (T1 plots)

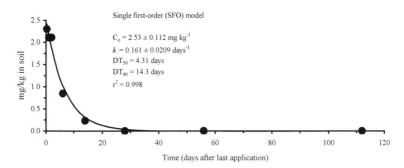

Single first-order (SFO) model

$C_o = 2.53 \pm 0.112$ mg kg^{-1}
$k = 0.161 \pm 0.0209$ days^{-1}
$DT_{50} = 4.31$ days
$DT_{90} = 14.3$ days
$r^2 = 0.998$

β-Endosulfan in soil (T1 plots)

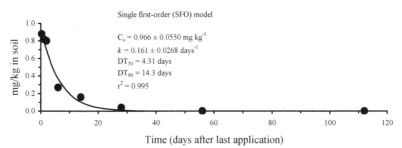

Single first-order (SFO) model

$C_o = 0.966 \pm 0.0550$ mg kg^{-1}
$k = 0.161 \pm 0.0268$ days^{-1}
$DT_{50} = 4.31$ days
$DT_{90} = 14.3$ days
$r^2 = 0.995$

Total endosulfan in soil (T1 plots)

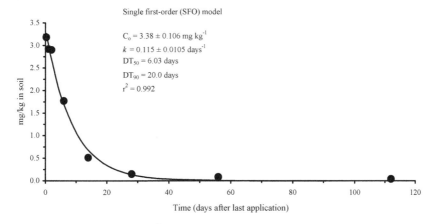

Single first-order (SFO) model

$C_o = 3.38 \pm 0.106$ mg kg^{-1}
$k = 0.115 \pm 0.0105$ days^{-1}
$DT_{50} = 6.03$ days
$DT_{90} = 20.0$ days
$r^2 = 0.992$

Figure 3.4. Monophasic dissipation of (a, top) α-endosulfan, (b, middle) β-endosulfan, and
(c, bottom) total endosulfan residues in tomato cropped soil

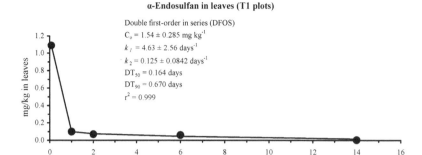

α-Endosulfan in leaves (T1 plots)

Double first-order in series (DFOS)
$C_o = 1.54 \pm 0.285$ mg kg^{-1}
$k_1 = 4.63 \pm 2.56$ days^{-1}
$k_2 = 0.125 \pm 0.0842$ days^{-1}
$DT_{50} = 0.164$ days
$DT_{90} = 0.670$ days
$r^2 = 0.999$

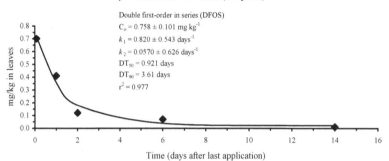

β-Endosulfan in leaves (T1 plots)

Double first-order in series (DFOS)
$C_o = 0.758 \pm 0.101$ mg kg^{-1}
$k_1 = 0.820 \pm 0.543$ days^{-1}
$k_2 = 0.0570 \pm 0.626$ days^{-1}
$DT_{50} = 0.921$ days
$DT_{90} = 3.61$ days
$r^2 = 0.977$

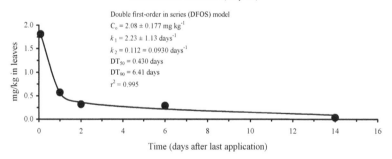

Total endosulfan in leaves (T1 plots)

Double first-order in series (DFOS) model
$C_o = 2.08 \pm 0.177$ mg kg^{-1}
$k_1 = 2.23 \pm 1.13$ days^{-1}
$k_2 = 0.112 = 0.0930$ days^{-1}
$DT_{50} = 0.430$ days
$DT_{90} = 6.41$ days
$r^2 = 0.995$

Figure 3.5. Biphasic dissipation of (a, top) α-endosulfan, (b, middle) β-endosulfan, and (c, bottom) total endosulfan residues in tomato foliage.

2 weeks after the last spraying (74% in plant and 26% in soil). We suggest that the dissipation occurred through volatilization and degradation of the pesticide in either plants or soil microorganisms.   In the study, there was no significant off-site movement of in-furrow irrigation water.   Additionally, there was only one small rainfall (11 mm) during the study. Therefore, there was little potential for endosulfan foliar wash-off. As has been discussed in previous sections, the dissipation of endosulfan in foliage is characterized by an initial rapid volatilisation phase. In our investigation, endosulfan loss through volatilisation was not measured.

Several authors (Farmer *et al.*, 1972; Spencer *et al.*, 1973; Kennedy *et al.*, 2001) have held the concept that volatilisation is a significant route of pesticide loss in the field, particularly when it is applied to the surfaces of soils or plants (Roger and Bhuiyan, 1995), and this may explain how traces of endosulfan have been found in areas never sprayed such as the Artic (Jantunen and Bidleman, 1998) and remote areas around the world (Simonich and Hites, 1995). In a study (Kennedy *et al.*, 2001) of the fate and transport of endosulfan in an Australian cotton field, the authors found approximately 70% of endosulfan dissipating through volatilisation, with only a small percentage (8.5%) remaining on-field a month after four foliar applications of Thiodan ULV (total load of 3000 g of ai/ha). Further studies are needed to quantify volatilisation to estimate a total field balance.

Acknowledgements

We thank Ato (Akumadan) for permission to work on his land and for his assistance in the field. We acknowledge the Kinneret Limnological Laboratory, Migdal, Israel, for technical assistance in the use of GC.

References

Aguilera-del real, A.; Valverde-García, A.; Fernandez-Alba, A.R.; Camacho-Ferre, F. Behaviour of endosulfan residues in peppers, cucumbers and cherry tomatoes grown in Greenhouse: Evaluation by decline curves. *Pestic. Sci.* **1997**, 51, 194-200.

Amoah, P.; Drechsel, P.; Abaidoo, R.C.; Ntow, W.J. Pesticide and Pathogen contamination of vegetables in Ghana's urban markets. *Arch. Environ. Contam.Toxicol.* **2006**, 50, 1-6.

Antonious, G.F.; Byers, M.E. Fate and movement of endosulfan under field conditions. *Environ. Toxicol. Chem.* **1997**, 16, 644-649.

Antonious, G.F.; Byers, M.E.; Snyder, J.C. Residues and fate of endosulfan on field-grown pepper and tomato. *Pestic. Sci.* **1998**, 54, 61-67.

Beard, J.E.; Ware, G.W. Fate of endosulfan on plants and glass. *J. Agric. Food Chem.* 1969, 17, 216-220.

Chopra, N.M.; Mahfouz, A.M. Metabolism of endosulfan I, endosulfan II, and endosulfan sulfate in tobacco leaf. *J. Agric. Food Chem.* **1977**, 25, 32-36.

Day, P.R. Particle fractionation and particle-size analysis. In *Methods of soil analysis;* Black, C.A., Ed.; American Society of Agronomy: Madison, WI, **1965,** 545-567.

Donker, M.H.; Eijsackers, H.; Heimbach, F. Ecotoxicology of soil organisms. *SETAC Special Publication series* **1994**, chapter 14.

EXTOXNET. Extension Toxicology Network: Pesticides Information Profiles. Avaialble from URL: *http://extoxnet.orst.edu/pips/ghindex.html.* Accessed 16 February 2007.

FAO/IAEA *Organochlorine insecticides in African agroecosystems.* FAO, Geneva, **1997, 93-**230.

Farmer, W.J.; Igue, K.; Spencer, W.F.; Martin, J.P. Volatility of organochlorine insecticides from soil: Effects of concentration, temperature, air flow rates and vapour pressure. *Soil Sci. Soc. Amer. Proc.* **1972**, 36, 443-447.

Ghadiri, H.; Rose, C.W.; Connel, D.W. Controlled environment study of the degradation of endosulfan in soils. P. 583-588. In *Challenging the future*; Constable, G.A. and Forrester, N.W.; Ed.; Proc. 1st World Cotton Res. Conf., CSIRO, Canberra, Australia, **1994**, 583-588.

Goebel, H.; Gorbach, S.G.; Knauf, W.; Rimpau, R.H.; Huttenbach, H. Properties, effects, residues, and analytics of the insecticide endosulfan. *Residue Rev.*, **1982**, 83, 1-165.

Gonzalez, M.; Miglioranza, K.S.B.; Aizpún de Moreno, J.E.; Moreno, V.J. Occurrence and distribution of organochlorine pesticides (OCPs) in tomato (*Lycopersicum esculentum*) crops from organic production. *J. Agric. Food Chem.* **2003**, 51, 1353-1359.

Guerin, F.G.; Kennedy, I.V. Distribution and dissipation of endosulfan and related cyclodienes in sterile aqueous systems: Implications for studies on biodegradation. *J. Agric. Food*

Chem. **1992**, 40, 2315-2323.

Jackson, M.L. Soil chemical analysis. Prentice Hall, Englewood Cliffs, NJ, **1958**.

Jantunen, L.M.M.; Bidleman, T.F. Organochlorine pesticides and enantiomers of chiral pesticides in Artic Ocean water. *Arch. Environ. Contam. Toxicol.* **1998**, 35, 218-228.

Kathpal, T.S.; Singh, A.; Dhankhar, J.S.; Singh, G. Fate of endosulfan in cotton soil under subtropical conditions of Northern India. *Pestic. Sci.* **1997**, 50, 21-27.

Kennedy, I.R.; Sánchez-Bayo, F.; Kimber, S.W.; Hugo, L.; Ahmad, N. Off-site movement of endosulfan from irrigated cotton in New South Wales. *J. Environ. Qual.* **2001**, 30, 683-696.

Kimber, S.W.L.; Coleman, S.; Coldwell, R.L.; Kennedy, I.R. The environmental fate of endosulfan sprayed on cotton. *Eighth Internat. Cong. Pest. Chem (IUPAC),* Washington, DC, USA, 4-9 July, Vol. 1, Abstract No. 234, **1994**, p.263.

Maier-Bode, H. Properties, effect, residues and analytics of the insecticide endosulfan. *Residue Rev.* **1968**, 22, 2-44.

Martens, R. Degradation of endosulfan-8, 9-^{14}C in soil under different conditions. *Bull. Environ. Contam. Toxicol.* **1977**, 17, 438-446.

Miglioranza, K.S.B.; Aizpún de Moreno, J.E.; Moreno, V.J.; Osterrieth, M.L.; Escalante, A.H. Fate of organochlorine pesticides in soils and terrestrial biota of "Los Padres" pond watershed, Argentina. *Environ. Pollut.* **199**, 105, 91-99.

Miles, J.R.W.; Moy, P. Degradation of endosulfan and its metabolites by a mixed culture of soil microorganisms. *Bull. Environ. Contam. Toxicol.* **1979**, 23, 013-019.

Morrica, P.; Fidente, P.; Seccia, S.; Ventriglia, M. Degradation of imazosulfuron in different soils-HPLC determination. *Biomed. Chromatogr.* **2002**, 16, 489-494.

Ntow, W.J. Organochlorine Pesticides in Water, Sediment, Crops and Human fluids in a farming community in Ghana. *Arch. Environ. Contam. Toxicol.,* **2001**, 40, 557-563.

Ntow, W.J. Pesticide residues in Volta Lake, Ghana. *Lakes & Reservoirs: Research and Management* **2005**, 10, 243-248.

Ntow, W.J.; Gijzen, H.J.; Kelderman, P.; Drechsel, P. Farmer perceptions and pesticide use practices in vegetable production in Ghana. *Pest Manag. Sci.* **2006**, 62, 356-365.

Osafo, S.; Frimpong, E. Lindane and endosulfan residues in water and fish in the Ashanti Region of Ghana. *J. Ghana Sci. Assoc.* **1998**, 1, 135-140.

Raha, P.; Banejee, H.; Das, A.K.; Adityachaudhury, N. Persistence Kinetics of Endosulfan, Fenvalerate, and Decamethrin in and on Eggplant (*Solanum melongena* L.). *J. Agric. Food Chem.* **1993**, 41, 923-928.

Roger, P.A.; Bhuiyan, S.I. Behavior of pesticides in rice-based agroecosystems: A review. In *Impact of pesticides on farmer health and the rice environment*; Pingali, P.L. and Roger, P.A. Eds.; Kluwer Academic Publishers, Boston, USA, **1995**, 111-148.

Schuphan, I.; Ballschmiter, K.; Tolg, G. Zum metabolismus des endosulfans in ratten and mausen. *Z. Naturforsch. B* **1968**, 23, 701-706.

Sethunathan, N.; Megharaj, M.; Chen, Z.; Singh, N.; Kookana, R.S.; Naidu, R. Persistence of endosulfan and endosulfan sulfate in soil as affected by moisture regime and organic matter addition. *Bull. Environ. Contam. Toxicol.* **2002**, 68, 725-731.

Shetty, P.K.; Mitra, J.; Murthy, N.B.K.; Namitha, K.K.; Savitha, K.N.; Raghu, K. Biodegradation of cyclodiene insecticide endosulfan by *Mucor thermohyalospora* MTCC 1384. *Curr. Sci.* **2000**, 79, 1381-1383.

Simonich, S.L.; Hites, R.A. Global distribution of persistent organochlorine compounds. *Science* **1995**, 269, 1851-1854.

Singh, P.P.; Battu, R.S.; Singh, B.; Kalra, R.L. Fate and interconversion of endosulfan I, II and sulfate on gram crop (*Cicer arietinum* Linn.) in subtropical environment. *Bull. Environ. Contam. Toxicol.* **1991**, 47, 711-716.

Spencer, W.F.; Farmer, W.J.; Claith, M.M. Pesticide volatilisation. *Residue Rev.* **1973**. 49, 1-47.

Stewart, D.K.R.; Cairns, K.G. Endosulfan persistence in soil and uptake by potato tubers. *J. Agric. Food Chem.* **1974**, 22, 984-986.

U.S.Department of Health and Human Services. Toxicological profile for endosulfan. Agency for toxic substance and disease registry, Atlanta. **1990**.

Chapter 4

The impact of agricultural runoff on the quality of two streams at Akumadan and Tono

Submission to a scientific journal based on this chapter:

William Joseph Ntow, Pay Drechsel, Benjamin O. Botwe, Peter Kelderman, and Huub J. Gijzen. The impact of agricultural runoff on the quality of two streams in vegetable farm areas in Ghana. *J. Environ. Qual.* Accepted for publication on October 17, 2007.

The impact of agricultural runoff on the quality of two streams at Akumadan and Tono

Abstract

A study of two small streams at Akumadan and Tono, Ghana, was undertaken during the rain and dry season periods between February 2005 and January 2006 in order to investigate the impact of vegetable field runoff on their quality. In each stream we compared the concentration of current-use pesticides in one site immediately upstream of a vegetable field with a second site immediately downstream. Only trace concentrations of endosulfan and chlorpyrifos were detected at both sites in both streams in the dry season. In the wet season, rain-induced runoff transported pesticides into downstream stretches of the streams. Average peak levels in the streams themselves were 0.07 µg/L endosulfan, 0.02 µg/L chlorpyrifos (the Akumadan stream); 0.04 µg/L endosulfan, 0.02 µg/L chlorpyrifos (the Tono stream). Respective average pesticide levels associated with streambed sediment were 1.34, 0.32 $\mu g/kg^{-1}$ (the Akumadan stream) and 0.92, 0.84 $\mu g/kg^{-1}$ (the Tono stream). Further investigations are needed to establish the potential endosulfan and chloypyrifos effects on aquatic invertebrate and fish in these streams. Meanwhile measures should be undertaken to reduce the input of these chemicals via runoff.

Keywords: non-point-source pollution; pesticides; runoff; sediment; streams; vegetable; water quality

Introduction

Non-point source agricultural pollution is regarded as the greatest threat to the quality of surface waters in rural areas. One of the most important routes leading to non-point source agricultural pollution of surface waters in rural areas is runoff. Runoff from agricultural fields introduces pesticides, soil, organic matter, manure and fertilizer into small streams, increasing the volume of stream discharge and changing water quality (Neumann and Dudgeon, 2002). The impacts of such runoff are well documented (Cooper, 1993; Castillo et al., 1997; DeLorenzo et al., 2001). Intensive vegetable farm areas of Akumadan and Tono, Ghana (Figure 4.1) provide an opportunity to study the effect of agriculture runoff on current-use pesticides levels of small streams in the tropics.

Vegetable production in Ghana typically occurs in intensely managed smallholder farms or irrigation schemes with relatively high inputs of pesticides. Recorded vegetable crops (tomato, pepper, okra, eggplants (or garden eggs), and onion) cover approximately 0.4% of the cultured land of Ghana, equating to 58,270 ha in 1998 (Gerken et al., 2001). An estimated average pesticide rate of 0.08 liters ai/ha is applied to these vegetables. Relatively lower quantities are applied to cereals (0.03 liters ai/ha[1]) and higher quantities to cocoa (0.5 liters ai/ha). The compounds applied in vegetable production include organochlorine and organophosphate insecticides. In contrast to the traditional organochlorines, organophosphates are not highly persistent, but some can be highly toxic to aquatic organisms (Castillo et al., 2006). Intensively managed vegetable farms are also characterized by an extensive network of drainage systems where surplus water may flow into local streams and rivers. Consequently, the aquatic ecosystems located downstream of vegetable farmlands might be vulnerable due to intensive pesticide use, drainage systems, and high precipitation rates typical for tropical areas where vegetable production occurs.

Although few studies (Osafo and Frimpong, 1998; Ntow, 2001) have dealt with pesticide residue levels in river ecosystems in farmlands of Ghana, little is known about transport pathways such as rain-induced runoff. Vegetable crops are sprayed with a range of chemical pesticides (Ntow *et al.*, 2006), and fields are cultivated up to the margins of streams where agriculture is practiced. Because of current land tenure systems which support land rotation (Gerken *et al.*, 2001), much agricultural land in Ghana has been abandoned (left to fallow). Active and abandoned agricultural land is generally situated in drainage basins. In this study, we investigated the effects of rain-induced runoff from vegetable fields on water quality by comparing the concentrations of current-use pesticides of paired sites upstream and downstream of vegetable farms along two streams, the Akumadan and the Tono. These sites were sampled at the end of the dry season and again at the start of the wet season when the streams received runoff after rainfall. Our null hypothesis was that the magnitude of the difference in levels of current-use pesticides between the upstream and downstream pairs of sites would remain unchanged between dry and wet season sampling. It was anticipated that any difference arising in the data set would manifest during the wet season, when high levels of current-use pesticides might occur downstream of vegetable field.

Materials and Methods

Study area
We selected for study two small streams that flow through areas of intensive cultivation of vegetables. The Akumadan stream is located near the village Akumadan, a prominent tomato-cultivating village in Offinso District in Ashanti Region (Figure 4.1). The surrounding areas of the Akumadan stream are characterized by intensive vegetable farming, mostly mixed cropping. Among the major crops cultivated are pepper, eggplants, okra, and tomatoes. Of the vegetables cultivated, tomatoes alone constitute over 90%. The tomato season runs through the whole year in four sub-seasons (Ntow, 2001). The area is subject to rain events of 150 rainy days/year or more, and an annual precipitation average of 1400 mm (Nurah, 1999). The predominant active ingredients of pesticides used at Akumadan are endosulfan and chlorpyrifos. Applications on vegetable fields occur on calendar basis (Ntow *et al.*, 2006), at approximately 7-day intervals, and at average rates of 1.0 and 0.04 kg ai/ha for endosulfan and chlorpyrifos, respectively.

The Tono stream is located near the Tono Irrigation Project, at Tono near Navrongo in Kassena-Nankana District in Upper East Region (Figure 4.1). The Tono Irrigation Project, under the management of Irrigation Company of the Upper Region (ICOUR), was started over a decade ago to promote the production of food crops by small-scale farmers within organized and managed irrigation schemes. The project covering 2,490 ha and divided amongst 3,000 farmers uses the waters of the Tono stream (Figure 4.1) for the purpose. The cropping areas are divided between upland and lowland areas on a ratio of 50:50. Crops grown in upland plots include onions, tomatoes, millet, groundnuts, sorghum and maize. The lowland areas have been developed for rice production. The number of rainy days per year is < 70 with an annual precipitation average of about 1000 mm. Predominant active ingredients in use in the area are endosulfan and chlorpyrifos. Applications on vegetable fields occur at average rates of 0.04 and 0.6 kgai/ha for endosulfan and chlorpyrifos, respectively, and when pests appear on crops. The applications are supervised and moderated by ICOUR. The physicochemical properties of endosulfan and chlorpyrifos are given in Table 4.1.

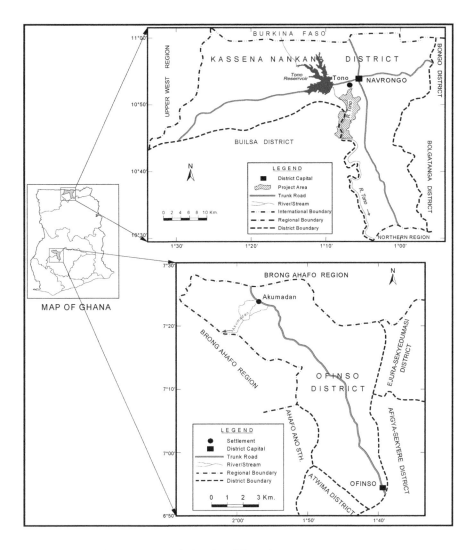

Figure 4.1. Map of Ghana showing the location of the study areas

Table 4.1. Physicochemical properties (EXTOXNET, 2006) of the predominantly used
pesticides in the catchments of the Akumadan and Tono streams

	Endosulfan	Chlorpyrifos
Water solubility (mg/L)	0.32 (22 $^{\circ}$C)	1.2 (25 $^{\circ}$C)
Soil half life, DT_{50soil} (days)	50	35-78
Soil adsorption coefficient, K_{oc}	12,400	6,070

Sampling procedure

Standard sampling procedure before and after the rainfall event included measurement of turbidity as total suspended solids (TSS) and collecting of water and streambed sediment for pesticide residue analysis at all sites. We sampled upstream and downstream sites in both streams (the Akumadan and the Tono) twice in 2005: in March and April (the Akumadan stream) and March and May (the Tono stream). Three replicate water and sediment samples were collected at each site on each date. In each stream an upstream site surrounded by abandoned fields was compared to a site downstream of a vegetable field where there is an agricultural activity. Apart from this difference, sites were selected to be similar in physical aspect, riparian features and substrate. To minimize the confounding effects of longitudinal variation, the upstream and the downstream sampling points were not more than 500 m apart. No point sources of pollution were evident within the study reaches. The first sample date for both sites corresponded to the end of the dry season in March when rain was infrequent. The second set of samples was taken just after rainfall (11 mm on April 11, 2005, the Akumadan stream, and 9 mm on May 8, 2005, the Tono stream) had caused runoff from the fields.

Water samples representing the pesticide levels during runoff were collected using a procedure described in Dabrowski *et al.* (2002a). Briefly, glass-sampling bottles were stored in the stream with the opening approximately 3 cm above the normal water level. During rainfall induced surface runoff, the rising water level filled the bottles passively. A small glass pipe tied in the opening of the bottle enabled a free flow of water into the bottle while air could flow out via the glass pipe. Retrieval of water samples took place within 24 h of the runoff event. Samples of pesticides associated with streambed sediments were collected manually with a stainless steel spoon from the top 1 cm surface layer into aluminum foils. All the samples were transported to the CSIR Water Research Institute Laboratory within 24 to 48 h on ice in clean ice-chests and stored in the laboratory refrigerator at 4 °C until time of analyses.

TSS was measured using a turbidity meter (2100P Turbidimeter, Hach Company, Loveland, CO, USA). To calibrate the turbidity measurements as described by Schulz (2001), certain samples were filtered through pre-weighed Whatman GF/F (0.45 μm pore-size) glass microfibre filters and dried at 60 °C for 48 h. The filter paper was then re-weighed to determine TSS.

Pesticide analysis

The extraction and analyses of water were performed following the Association of Official Analytical Chemists 990.06 and 970.52 methods (Rovedatti *et al.*, 2001). Briefly, the 1.0 L unfiltered water samples were extracted sequentially three times with 25 mL *n*-hexane each time. The extract was dried with anhydrous sodium sulfate and concentrated down to 10 mL by evaporation in a TurboVap (Zymark, Palo Alto, California, USA). A clean-up system, using a chromatographic column packed with florisil, previously activated for 3 h in an oven at 130 °C, and anhydrous sulfate (both rinsed with petroleum ether) was used. The extract was transferred to the column. Three fractions were obtained after elution with 6, 15, 50% ethyl ether in petroleum ether. Maximal flux rate of elution was 5 mL/min. Each eluate was evaporated. The extracts were dissolved in 1.5 mL *n*-hexane and made up to 2 mL with more *n*-hexane. The dissolved extracts were injected into a gas chromatographic system for identification and quantification of the pesticides.

Sediment samples were well mixed to obtain a homogeneous sample and then transferred into pans to air-dry at ambient temperature. The air-dried sediment samples were ground in a mortar and sieved (2 mm). The extraction procedures are described by Ntow (2001). Briefly, about 5g representative sieved samples were weighed into extraction thimbles and soxhlet-extracted in methanol and cleaned up in florisil as described above for water.

Gas chromatography was performed with an Agilent 6890 coupled with Agilent 5973N mass selective detector-electron impact ionization. The capillary column was HP-5MS (length 30 m; I.D 0.25 mm and film thickness 0.25 μm) and packed with 5% phenyl methyl siloxane. The GC-MS was operated in the selected ion-monitoring mode at the following conditions: injection port 250 °C (splitless, pressure 22.62 psi; purge flow 50 mL/min; purge time 2.0 min; total flow 55.4 mL/min). Column oven: initial 70 °C, held 2 min, programming rate 25 °C min^{-1} (70 to 150 °C); 10 °C min^{-1} (150 to 200 °C); 8 °C/min (200 to 280 °C) and held 10 min at 280 °C. The carrier gas was nitrogen at 15 psi. The injection volume was 1 μL (Agilent 7683 Series injector).

Analyte recovery experiments were performed with the water and sediment matrices. One-liter samples of distilled tap water were spiked with 0.01 μg of each pesticide standard. The samples were then extracted and analyzed, in accordance with the previously noted procedure. Similarly, uncontaminated sediment (taken from the premises of the CSIR Water Research Institute, Accra, Ghana) was spiked with known quantities of pesticides before extraction (0.1 μg/substance in 5 g of sediment dry mass) and was then processed and analyzed as described previously. Recovery of the different pesticides ranged between 79% and 104% with the variation coefficients not exceeding 13%. The concentration estimates were not corrected for these recoveries. The following quantification limits were obtained for water and sediment: 0.01 μg/L and 0.05μg/kg dry weight, respectively (calculated from real samples as being 10 times the signal: noise ratio). Three replicates of samples were used. During the sample extraction, blanks were regularly processed (one in ten). The standard deviation of the pesticides concentrations in three equal samples ranged between 2.9 and 6.1% depending on the sample and on the compound. The screening included the following pesticides: (a) organochlorines: α- and β-endosulfan, endosulfan sulfate (b) organophosphates: chlorpyrifos. Selection of analyzed pesticides was done on the basis of use information gathered during a survey.

Differences in concentration of pesticides between upstream and downstream sites on each sampling date were analyzed by one-way-analysis of variance followed by a Bonferroni test (equal variances assumed) (SPSS software, version 12.0.1 for Windows, SPSS Inc, Chicago, Illinois, USA). The data were analyzed for normal distribution (Kolmogorov-Smirnov test) and homogeneity of variance (Levene's Test) prior to statistical analysis.

Results

Pesticides in water and sediment
Pesticides measurements undertaken in dry and wet seasons are summarized in Table 4.2. In both streams, the Akumadan and the Tono, sites situated upstream and downstream of vegetable fields were free of current-use pesticide contamination in water samples taken in March, during the dry season period of investigation, considering a quantification limit of 0.01 μg/L. However, in the wet season in April and May, after runoff events, the pattern in both streams was changed.

Table 4.2. Mean concentration ($\pm$ SE; n = 3) of pesticides in water in Akumadan and Tono streams (μg/L). Statistical differences between sampling periods (ANOVA) are indicated.

Sampling period	Akumadan stream				Tono stream			
	Upstream		Downstream		Upstream		Downstream	
	Endosulfan [a]	Chlorpyrifos	Endosulfan [a]	Chlorpyrifos	Endosulfan [a]	Chlorpyrifos	Endosulfan [a]	Chlorpyrifos
Dry season	nd [b]	nd [b]	nd [b]	nd [b]	nd [b]	nd [b]	nd [b]	nd [b]
Wet season	0.01 ± 0.00	nd [b]	0.07 ± 0.01	0.02 ± 0.00	nd [b]	0.01 ± 0.00	0.04 ± 0.01	0.02 ± 0.00
p-value [c]	*	ns	***	***	ns	*	***	***

[a] = α-endosulfan + β-endosulfan + endosulfan sulfate; = below the quantification limit of 0.01 μg/L ; [c] = * $p \leq 0.05$; ** $p \leq 0.01$; *** $p \leq 0.001$; ns = not significant

Table 4.3. Mean concentration ($\pm$ SE; n = 3) of pesticides in streambed sediment (μg/kg) of Akumadan and Tono streams. Statistical differences between sampling periods (ANOVA) are indicated.

Sampling period	Akumadan stream				Tono stream			
	Upstream		Downstream		Upstream		Downstream	
	Endosulfan [a]	Chlorpyrifos	Endosulfan [a]	Chlorpyrifos	Endosulfan [a]	Chlorpyrifos	Endosulfan [a]	Chlorpyrifos
Dry season	0.42 ± 0.01	0.05 ± 0.01	0.41 ± 0.01	0.06 ± 0.03	0.36 ± 0.01	0.19 ± 0.02	0.35 ± 0.01	0.16 ± 0.01
Wet season	0.49 ± 0.02	0.05 ± 0.02	1.34 ± 0.04	0.32 ± 0.05	0.36 ± 0.01	0.15 ± 0.01	0.92 ± 0.01	0.84 ± 0.07
p-value [b]	*	ns	***	***	ns	*	***	***

[a] = α-endosulfan + β-endosulfan + endosulfan sulfate; [b] = * $p \leq 0.05$; ** $p \leq 0.01$; *** $p \leq 0.001$; ns = not significant

Water samples taken during the wet season under runoff conditions showed no or, at the most, very low (0.01 µg/L) pesticide contamination of upstream sites. Relatively, increased water contaminations by endosulfan and chlorpyrifos, 0.07 and 0.02 µg/L, respectively, for the Akumadan stream and 0.04 and 0.02 µg/L, respectively, for the Tono stream, were detected at downstream sites in the corresponding samples for the wet season (Table 4.2). Thus, regarding the downstream sections of the two streams, the Akumadan and the Tono, pesticides concentrations measured in the wet season were increased significantly ($p \leq 0.001$) above the corresponding concentrations in the dry season in March. Additionally, while there were only minor differences in pesticides levels in water samples collected in dry versus wet season in upstream sites of the streams; the respective differences were very significant ($p \leq 0.001$) in downstream sites.

The mean concentrations of both endosulfan and chlorpyrifos in sediment did not differ ($p > 0.001$) between dry (0.42 and 0.05 µg/kg, respectively) and wet (0.49 and 0.05 µg/kg respectively) seasons in upstream sections of the Akumadan stream (Table 4.3). A similar pattern was observed for the Tono stream (dry season: 0.36 µg/kg endosulfan, 0.19 µg/kg chlorpyrifos; wet season: 0.36 µg/kg endosulfan, 0.15 µg/kg chlorpyrifos) in the upstream stretch. However, with the onset of rains, the situation changed markedly in both streams. While the upstream sites did not change significantly ($p > 0.001$) in concentration of pesticides between the sampling periods, there were significant ($p \leq 0.001$) changes in concentration in the downstream sections of both streams. For instance, for the same sampling site (downstream section of the Akumadan stream), endosulfan concentration in the sediment increased from 0.41 to 1.34 µg/kg; chlorpyrifos from 0.06 to 0.32 µg/kg, respectively, after the storm event, due to runoff from the vegetable field. Similarly, for the Tono stream, endosulfan concentration in the sediment increased from 0.35 to 0.92 µg/kg; chlorpyrifos from 0.16 to 0.84 µg/kg, respectively (Table 4.3).

Comparing the water and sediment samples, generally, the sediment samples exhibited the highest concentrations of pesticides. For instance, endosulfan concentrations increased to 0.49 and 1.34 µg/kg in streambed sediment in the Akumadan stream at upstream and downstream sites, respectively, in the wet season, which are equivalent to increases by factors of 49 and 19, respectively, in comparison to the water phase. Similarly, for endosulfan in the Tono stream, an increase by a factor of 23 was observed at downstream site for the wet season sampling period. Chlorpyrifos also exhibited comparable factors as for endosulfan for the water and sediment phases in the two streams in the wet season.

To further assess the effect of vegetable field runoff, we calculated the difference between the mean concentration of pesticides at the downstream and upstream sites on each stream on both sampling dates. To account for residues concentrations that were below the limit of quantification, we assumed that actual concentrations for non-detect (nd) samples were equal to the limit of quantification. However, this would underestimate the differences if there would be no contamination of the sampled water and sediment. On the other hand, since the comparison here was relative, not absolute, size of the effect, it was decided to stick to this assumption. The concentration of pesticides was always higher at the downstream sampling points during the wet season (Figure 4.2). The effect was particularly strong for endosulfan and chlorpyrifos in water and sediment, respectively, for both streams.

Frequency of occurrence of residues of endosulfan isomers and sulfate metabolite was in general higher for both water and streambed sediment of downstream sites of the two streams. α-Endosulfan and β-endosulfan exhibited frequencies of detections of approximately 90% in downstream sites and 40-60% in upstream sites, respectively. Approximately, 70% of the total endosulfan detected in the two streams was identified as the α- and β-isomers (Table 4.4).

Water

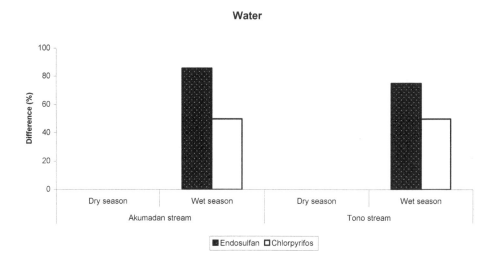

Sediment

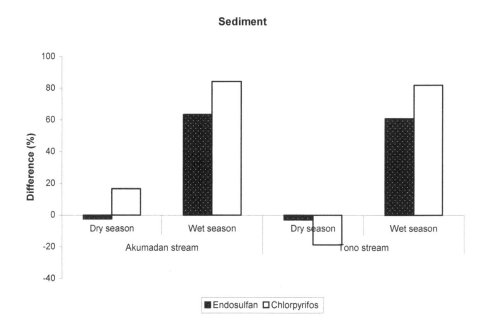

Figure 4.2. Percentage difference between mean concentrations of pesticides at upstream and downstream sampling points on the Akumadan and the Tono streams. The data are presented separately for water and sediment (see Tables 4.2 and 4.3).

Table 4.4. Detected endosulfan residues with their relative abundance and the frequency of occurrence (*n*) at 24 sampling sites (i.e., 3 composite samples times 1 stream site times 2 streams times 2 sample types times 2 sampling occasions).

Endosulfan residues	Relative abundance in all samples combined (%)	Upstream Frequency of occurrence (*n*/24)	Downstream Frequency of occurrence (*n*/24)
α-Endosulfan	66.7	11	21
β-Endosulfan	72.9	13	22
Endosulfan sulfate	39.6	10	9

Total suspended solids
Turbidity measurements were taken before and after runoff events (within 24 h) and an average (of three samples) was obtained for each sampling site. After runoff, the TSS levels were considerably increased at all sites relative to the pre-runoff conditions (Figure 4.3). During normal flow conditions before runoff (in the dry season), all of the sites had TSS levels of about or less than 10 mg/L. The TSS levels were increased to about 20 mg/L (upstream sites) and 30-40 mg/L (downstream sites), which were approximately 2-4 fold higher after the rainfall event than the average dry season values.

Discussion

The results of our investigation show that vegetable field runoff had a relative effect on current-use pesticides content of two Ghanaian streams. This effect was evident from a comparison of upstream and downstream sites in the streams before and after rainfall events and consequent runoff from vegetable fields, which impacted the downstream sites. Both streams showed a relative downstream increase in concentration of current-use pesticides after the runoff event. Vegetable fields dominated all of the downstream sub-catchments. On the contrary, runoff did not result in any significant increased contamination in upstream sites where no vegetable fields were present. The relatively high concentrations of pesticides detected in both the water (dissolved + adsorbed) and streambed sediment phases in downstream sub-catchments of the streams in the wet season sampling highlight the impact that runoff from vegetable fields can have on a small stream. Upstream and downstream sampling sites were chosen close together, eliminating the effect of longitudinal variation. Point-source discharge of pesticides was unlikely to have been a confounding factor as this was not evident within the study reaches.

There is much evidence to suggest that measured pesticide concentrations were as a result of surface runoff and not via alternate routes such as rapid leaching through the soil substrate (Dabrowski *et al.*, 2002a). Endosulfan and chlorpyrifos were relatively prevalent in downstream water (dissolved + adsorbed) and sediment samples collected during runoff. Both of these pesticides have a high affinity to adsorb to soil particles (Table 4.1) and are thus relatively immobile in soils (EXTOXNET, 2006). Thus, the most feasible way that these particles could land up in the streams is via surface runoff. The relative increase in turbidity (suspended sediment; Figure 4.3) after the storm events is another factor that suggested that runoff had taken place (Dabrowski *et al.*, 2002a). The increases in turbidity must have been as a result of soil and sediment being physically washed into the streams, transporting adsorbed pesticides in the process, and are unlikely to have occurred via sub-surface drainage.

The proportion of endosulfan isomers in total endosulfan detected in the study gave

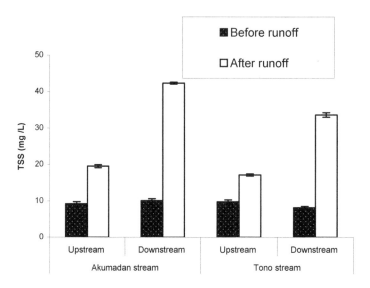

Figure 4.3. Mean (± SE; n = 3) total suspended solids (TSS) levels before and during runoff
conditions in Akumadan and Tono streams.

current-use pesticides applied in vegetable production in the catchments. Commercial
formulations of endosulfan (e.g. Thionex, Thiodan) used in vegetable production in Ghana
(Ntow *et al.*, 2006) contained two isomers, α- and β-isomers, with a higher relative amount
of α-isomer than β-isomer (α-isomer:β-isomer = 1.96:1). According to Kimber *et al.*
(1994), Kathpal *et al.* (1997) and Antonious *et al.* (1998), although the endosulfan α-
isomer is about 70% of the active ingredient in commercial formulations, it is found in
aquatic environments and solid surfaces at appreciable levels only immediately after
spraying, due to its high volatility. Endosulfan α-isomer is more volatile (vp = 0.006 mm
Hg at 20 °C) and less water-soluble (2.29 mg/L at 22 °C) compared to the β-isomer (vp =
0.003 mm Hg at 20°C and water solubility = 31.1 mg/L at 22 °C) (Guerin and Kennedy,
1992; Antonious *et al.*, 1998). Approximately, 70% of the total endosulfan detected in the
present study was identified as the α- and β-isomers, and can therefore be attributed to
recently applied chemical instead of residual concentrations.

Pesticide analysis in this study was carried out on unfiltered water samples. The
relative importance of pesticide transport dissolved in water or adsorbed onto suspended
solids has been investigated by several authors on a range of pesticides and reported in
Kreuger (1998). These results suggest that the greater part of the pesticide load is carried
dissolved in water. Wauchope (1978) summarized several runoff studies and concluded
that most of the pesticide contamination is introduced in the water phase, not necessarily
because the concentration is higher there but rather because the runoff contains
considerably more water than solid components. Nevertheless, the distribution of any
substance in the water or solid phase of the runoff depends mainly on its water solubility.
According to Wauchope (1978) the importance of particle-bound inputs increases for
insecticides with water solubility above 1 mg/L. Although the partitioning behavior of
endosulfan and chlorpyrifos is in general accordance with their physicochemical properties
(Table 4.1), both chemicals have been found at relatively high concentrations in matrices,
water and suspended particles (Dabrowski *et al.*, 2002a). Therefore, it is reasonable to
suppose that, as in the present investigation, the absolute pesticide amounts are generally
in both the water and adsorbed phases.

The runoff sampling method used in the present study represents a reasonable technique for sampling in a challenging environment. It is comparatively cheap and easy to install. Reliable qualitative information about the pesticide contamination can be obtained by use of this sampling method. A distinct disadvantage of the method is that it is not suitable for quantitative measurements of chemical transport. In many cases, however, it is more important to know whether any input of pesticides into a stream has occurred, and the method can provide an answer to this question. The equipment for quantitative, timed water/sediment sampling is expensive and usually time-consuming to operate (Schulz et al., 1998). The qualitative measurement offers the opportunity to obtain real contamination data from the field without the necessity of measuring any other parameters, as for a deterministic model (Schulz et al., 1998).

Summarizing, agricultural field runoff includes pesticides (Wauchope, 1978; Schulz et al., 1998; Neumann and Dudgeon, 2002) and sediments (Schulz, 2001). Both may degrade the water quality, but the contamination is only transient after heavy rainfall. The potential of runoff as a route of entry for pesticides and sediments input from agricultural fields into streams has been illustrated in a number of other studies (e.g. Wauchope, 1978; Readman et al., 1992; DeLorenzo et al., 1999; McDonald et al., 1999; Werner et al., 2000; Schulz, 2001; Schulz et al., 2001; Dabrowski et al., 2002b). Schulz (2001) and Nakano et al. (2004) provide an overview of field studies undertaken in temperate latitudes that establish a relationship between runoff events and increased total suspended sediment and pesticide levels in a river by monitoring their residues in river water or sediments. In this study, we were able to illustrate similar increases in tropical streams by focusing on a comparison of upstream and downstream sites in the streams before and after rainfall events and consequent runoff from vegetable fields. Because the study streams did not receive point-source discharge of pesticides, the only feasible way that the pesticides could have entered the streams was via surface runoff from the vegetable fields. Accordingly, we conclude that it was runoff from the vegetable fields, which impacted the downstream sites of the streams. Differences in agricultural activity in the upstream and downstream sub-catchments could be invoked to the variation in pesticide contamination between the sampling sites.

Acknowledgements

The authors are grateful to the Dutch Government (through the UNESCO-IHE Institute for Water Education), the International Water Management Institute (IWMI) and the International Foundation for Science (IFS) for their financial support of this research. The Kinneret Limnological Laboratory, Migdal, Israel, is acknowledged for technical assistance in the use of GC/MS.

References

Antonious, G.F., M.E. Byers, and J.C. Snyder. 1998. Residues and fate of endosulfan on field-grown pepper and tomato. Pestic. Sci. 54:61-67.

Castillo, L.E., E. De la Cruz and C. Ruepert. 1997. Ecotoxicology and pesticides in tropical aquatic ecosystems of Central America. Environ. Toxicol. Chem. 16:41-51.

Castillo, L.E., E. Martínez, C. Ruepert, C. Savage, M. Gilek, M. Pinnock, and E.Solis. 2006. Water quality and macroinvertebrate community response following pesticide applications in a banana plantation, Limon, Costa Rica. Science of the Total Environment 367:418-432.

Cooper, C.M. 1993. Biological effects of agriculturally derived surface-water pollutants on aquatic systems-a review. J. Environ. Qual. 22:402-408.

Dabrowski, J.M., S.K.C. Peall, A.J. Reinecke, M. Liess, and R. Schulz. 2002a. Runoff-related pesticide input into the Lourens River, South Africa: Basic data for exposure assessment and risk mitigation at the catchment scale. Water, Air, and Soil Pollution 135:265-283.

Dabrowski, J.M., S.K.C. Peall, A.V. Niekerk, A.J. Reinecke, J.A. Day, and R. Schulz. 2002b. Predicting runoff-induced pesticide input in agricultural sub-catchment surface water: linking catchment variables and contamination. Water Res. 36:4975-4984.

DeLorenzo, M.E., G.I. Scott, and P.E. Ross. 1999. Effects of the agricultural pesticides atrazine, deethylatrazine, endosulfan, and chlorpyrifos on an estuarine microbial food web. Environ. Toxicol. Chem.18:2824-2835.

DeLorenzo, M.E., G.I. Scott, and P.E. Ross. 2001. Toxicity of pesticides to aquatic microorganisms: A review. Environ. Toxicol. Chem. 20:84-98.

EXTOXNET. 2006. Extension Toxicology Network: Pesticides Information Profiles. Available from URL: http://extoxnet.orst.edu/pips/ghindex.html. Accessed 17 November 2006.

Gerken, A., J.V. Suglo, and M. Braun. 2001. Pesticide policy in Ghana. MoFA/PPRSD, ICP Project, Pesticide Policy Project/GTZ, Accra.

Guerin, F.G. and I.V. Kennedy. 1992. Distribution and dissipation of endosulfan and related cyclodienes in sterile aqueous systems: Implications for studies on biodegradation. J. Agric. Food Chem. 40:2315-2323.

Kathpal, T.S., A. Singh, J.S. Dhankhar, and G. Singh. 1997. Fate of endosulfan in cotton soil under sub-tropical conditions of Northern India. Pestic. Sci. 50:21-27.

Kimber, S.W.L., S. Coleman, R.L. Coldwell, and I.R. Kennedy. 1994. The environmental fate of endosulfan sprayed on cotton. Eighth Internat. Cong. Pest. Chem (IUPAC), Washington, DC, USA, 4-9 July, Vol. 1, Abstract No. 234, p.263.

Kreuger, J. 1998. Pesticides in stream water within an agricultural catchment in Southern Sweden, 1990-1996. The Science of the Total Environment 216:227-251.

McDonald, C., S.J. Jebellie, C.A. Madramootoo, and G.T. Dodds. 1999. Pesticide mobility on a hillside soil in St. Lucia. Agric. Ecosyst. Environ. 72:181-188.

Nakano, Y., A. Miyazaki, T. Yoshida, K. Ono, and T. Inoue. 2004. A study on pesticide runoff from paddy fields to a river in rural region-1: field survey of pesticide runoff in the Kozakura River, Japan. Water Res. 38:3017-3022.

Neumann, M. and Dudgeon D. 2002. The impact of agricultural runoff on stream benthos in Hong Kong, China. Water Res. 36:3103-3109.

Ntow, W.J. 2001. Organochlorine pesticides in water, sediment, crops and human fluids in a farming community in Ghana. Arch. Environ. Contam. Toxicol. 40:557-563.

Ntow, W.J., H.J. Gijzen, P. Kelderman, and P. Drechsel. 2006. Farmer perceptions and pesticide use practices in vegetable production in Ghana. Pest. Manag. Sci. 62:356-365.

Nurah, G.K. 1999. A baseline study of vegetable production in Ghana. National Agricultural Research Project (NARP) Report, Accra.

Osafo, S. and E. Frimpong. 1998. Lindane and endosulfan residues in water and fish in the Ashanti region of Ghana. J. Ghana Sci. Assoc. 1:135-140.

Readman, J.W., L. Liong Wee Kwong, L.D. Mee, J. Bartocii, G. Nilve, J.A. Rodríguez-Solano, and F. González-Farias. 1992. Persistent organophosphorus pesticides in tropical marine environments. Mar. Pollut. Bull. 24:398-402.

Rovedatti, M.G., P.M. Castañé, M.L. Topalián, and A. Salibián. 2001. Monitoring of organochlorine and organophosphorus pesticides in the water of the Reconquista River (Buenos Aires, Argentina). Water Res. 35:3457-3461.

Schulz, R. 2001. Rainfall-induced sediment and pesticide input from orchards into the Lourens River, Western Cape, South Africa: Importance of a single event. Water Res. 35:1869-1876.

Schulz, R., M. Hauschild, M. Ebeling, J. Nanko-Drees, J. Wogram, and M. Liess. 1998. A qualitative field method for monitoring pesticides in the edge-of-field runoff. Chemosphere 36:3071-3082.

Schulz, R., S.K.C. Peall, J.M. Dabrowski, and A.J. Reinecke. 2001. Spray deposition of two insecticides into surface waters in a South African orchard area. J. Environ. Qual. 30:814-822.

Wauchope, R.D. 1978. The pesticide content of surface water draining from agricultural fields-a review. J. Environ. Qual. 7:459-472.

Werner, I., L.A. Deanovic, V. Connor, V. DeVlaming, H.C. Bailey, and D.E. Hinton. 2000. Insecticide-caused toxicity to *Ceriodaphnia dubia* (Cladocera) in the Sacramento-San Joaquin River Delta, California, USA. Environ. Toxicol. Chem. 19:215-227.

Chapter 5

Occupational exposure to pesticides: blood cholinesterase activity in farmers at Akumadan

Submission to a scientific journal based on this chapter:

William J. Ntow, Laud M. Tagoe, Pay Drechsel, Peter Kelderman, and Huub J. Gijzen. Occupational exposure to pesticides: blood cholinesterase activity in a farming community in Ghana. *Arch. Environ. Contam. Toxicol.* Accepted for publication on October 20, 2007.

Occupational exposure to pesticides: blood cholinesterase activity in farmers at Akumadan

Abstract

A survey was undertaken to establish the extent of pesticide exposure in a farming community. Cholinesterase (ChE) activity in whole blood was used as a marker for assessing exposure to pesticides. Complete data were gathered for 63 farmers at Akumadan (exposed) and 58 control subjects at Tono, both prominent vegetable-farming communities in Ghana, by means of a questionnaire and blood cholinesterase analyses (acetylcholine-assay). Although whole blood ChE was significantly lower in the exposed than the control participants, it was not significantly correlated with either confounders of age, sex, body weight, and height or high risks practices. The high risks practices revealed during the survey included lack of use of personal protective clothing, short re-entry intervals, and wrong direction of spraying of pesticides by hand or knapsack sprayer. About 97% of exposed participants had experienced symptoms attributable to pesticide exposure. The frequent symptoms were reported as weakness and headache. There is the need to review safety precautions in the use and application of pesticides in Ghana.

Keywords: cholinesterase; carbamates; chlorpyrifos; organophosphates; pest control; pesticides

Introduction

The use of pesticides to control insect pests, which cause damage to crops and result in severe loss in food production in tropical countries like Ghana, has become recognised and accepted as an essential component of modern agricultural production. However, prolonged use of pesticides along with lack of suitable averting behaviour/use of basic protective requisites enhances the probability of accidental intoxication significantly. In Ghana, pesticides are typically applied to vegetables. Safety measures are generally poorly applied and workers lack proper knowledge or training in safe handling and application of these chemicals (Ntow *et al,.* 2006). These practices may produce a population with high exposures.

Two kinds of measurement: (1) enzyme activities in blood, and (2) unchanged pesticides and their metabolites in urine or blood have been used in biological monitoring for assessing exposure to pesticides. The assays of cholinesterase (ChE) activity in whole blood and erythrocytes are mainly applied to estimate inhibition by OPs and carbamates (WHO, 1986a; 1986b; He, 1993). However, under many field conditions procedures using whole blood are more practical than those using separated erythrocytes (He, 1993). Besides, since the serum and erythrocyte enzymes of whole human blood differ importantly, studies on either enzyme alone are inadequate to value the role played by the cholinesterase activity of whole blood (Alles and Hawes 1939 Downloaded from www.jbc.org on December 19, 2006).

Many organophosphate (OP) and carbamate pesticides are detrimental to non-target organisms such as aquatic organisms, birds, reptiles, and mammals (including humans). OPs and carbamates act as anti-cholinesterase (anti-ChE) agents. Because of the specificity of this relationship, depression of ChE activity has been used as a biomarker of exposure to these classes of pesticides with few reports on humans (Ohayo-Mitoko *et al.* 2000; Nielsen and Andersen 2002; Farahat *et al.,* 2003; Fenske *et al*., 2005). There is relatively little research on Ghanaian workers occupationally exposed to OP and carbamate pesticides. The objectives of this study were to determine the existing levels of whole blood ChE of vegetable farmers; evaluate associations between activities of exposed farmers that encourage high exposures and whole blood ChE; and describe the prevalence of symptoms attributable to pesticide exposure.

Materials and Methods

A sampling frame for this study was constructed from data available on vegetable production from the Ministry of Food and Agriculture (MoFA), crop protection from Plant Protection and Regulatory Services Directorate at MoFA, as well as information obtained from Nurah (1999). From this information, and for other practical reasons, it was decided to conduct the study in the vegetable farming community of Akumadan (see Figure 4.1, chapter 4). At Akumadan, the population of about 25,000 is engaged largely in vegetable farming. Among the major vegetables cultivated are pepper, garden eggs (eggplants), okra, and tomatoes. Of the vegetables cultivated, tomatoes alone constitute over 90%. The tomato season runs through the whole year (Ntow, 2001). At Akumadan, a wide and changing array of pesticides are utilised in the cultivation of vegetables. Table 5.1 lists types and trade names of pesticides used and their application. According to the severity of infestation, either one or a combination of two or three of the organochlorine (OC), OP, carbamate, or pyrethroid insecticides was applied during the crop-growing season. The pesticides were applied mostly by knapsack sprayers (portable backpack sprayer, carrying capacity 15 litre) and rarely by hand (brush, broom or leaves tied together to splash pesticides from a bucket). Sprayers were exposed by both inhalation and skin contact, and occasionally ingestion of contaminated food and/or drinking water.

The target population was farmers, of either gender, 18 years or above, from Akumadan working in vegetable farms, and applying pesticides during spray season (exposed). The principal investigator initially contacted the farmers through their chiefs, group leaders and/or fellow farmers to afford a forum to explain the purpose of the study. About 800 applicators of pesticides were initially recruited into the study. The control group consisted of farmers who had not handled pesticides in at least the preceding 2 months, who were initially selected from Tono Irrigation Project at Tono near Navrongo in Kassena-Nankana Disrict of Upper East Region of Ghana (Figure 4.1). About 2 months prior to the study, farmers at Tono Irrigation Project were not farming (and for that matter spraying pesticides) because the water pumping system was damaged. The ideal approach was to monitor changes in blood ChE activity of the study population during the agricultural season. Each farmer could then serve as his control. In practice, this was difficult to do since the farmers worked all through the year. According to Coye et al. (1986) a period of at least 30 days must separate the baseline determination from the farmer's last exposure to a cholinesterase-inhibiting compound. This was difficult to achieve due to the farmers' commitment to their duties. The mean whole blood ChE level obtained for the population at Tono Irrigation Project, therefore, served as a reference. Rama and Jaga (1992) have used literature values as a reference for their study on farm workers in the Republic of South Africa.

Table 5.1. Pesticides used for spraying vegetable crops at Akumadan, Ghana, in the period February 2005 to January 2006

Types and trade names of used pesticides	Active ingredient (ai)	Application rate* (kg ai/ha)	Number of applications*
Organochlorine compounds			
Thionex 35 EC/ULV	Endosulfan	0.04-1.0	6-12
Thiodan 50 EC	Endosulfan	0.02-0.8	6-12
Organophosphorus compounds			
Dursban 4E	Chlorpyrifos[‡]	0.04-0.6	6-12
Perferkthion 400 EC	Dimethoate[‡]	0.04-0.8	6-12
Carbamate compounds			
Furadan 3G	Carbofuran[‡]	0.06-0.9	6-12
Kontakt	Phenmedipham	0.06-0.8	6-12
Dithane M-45	Mancozeb	0.04-0.9	6-12
Trimangol 80	Maneb	0.04-0.8	6-12
Pyrethroids			
Karate 2.5 EC/ULV	Lambda cyhalothrin	0.01-0.2	6-12
Cyperdin	Cypermethrin	0.02-0.6	6-12
Cymbush	Cypermethrin	0.02-0.6	6-12
Decis	Deltamethrin	0.01-0.8	6-12

* Depending on the type of vegetable crop and also on farmer's knowledge
‡ Anti-cholinesterase agents

The estimated number of farmers using pesticides at Akumadan was 96% (Abraham Owusu, personal communication). The sample size was determined in order to have 95% confidence limits of 5% maximum error of the estimate, when the prevalence is 96% (Wadsworth Jr., 1990; Hogg and Tanis, 1997). This leads to a requirement of about 60 farmers for Akumadan. For a no-response expectation, the sample size was increased to 100 farmers. Thus, a sample of 100 farmers who represented a random sample from the population of exposed farmers (N = 800) was selected for the study. Similarly, 100 farmers from the population of control farmers from the Tono Irrigation Project were selected for the study.

All participants gave informed consent after receiving oral and/or written information on the project. The Ethics Review Committee of the Health Research Unit of the Ghana Health Service approved the study.

At midday from about 12 00 pm to 3 30 pm in the field, during which the farmers had rest and/or avoided exposure to the high temperature at that part of the day, blood was drawn with heparin-treated syringes and needles with the assistance of health workers and transferred to 5ml disposable plastic tubes already containing EDTA anti-coagulant. Samples were placed in an ice-cooled chest and delivered to the nearest hospital within 1 h for storage at –20 °C. Samples were later transported to the laboratory in Accra for analysis.

A Gallenkamp UV-Visible Spectrophotometer was used throughout the study for the measurement of ChE activity. The instrument was calibrated for wavelength and absorbance in accordance with the manufacturer's instructions. All the reagents used in the study were of analytical grade and used without any further purification. Acetylthiocholine iodide (ATC) and 5,5'-dithiobis-2-nitrobenzoic acid) (DTNB) were purchased from the Sigma-Aldrich chemical company, Germany.

DTNB reagent (10mmol/L): 61mL of 100mmol/L (14.2g/L) Na_2HPO_4 and 39 mL of 100mmol/L NaH_2PO_4 (13.8g/L) were mixed. pH of the resulting solution was checked (if too low dibasic salt was added to reach pH = 7 and if too high monobasic salt was added to reach pH = 7). 39.6mg of 5,5'-dithiobis-2-nitrobenzoic acid was dissolved in 10ml of the 100 mmol/L of sodium phosphate dibasic buffer. The reagent was stored in a dark bottle at 4 °C.

ATC substrate (75mmol/L): 5 mL of 100mmol NaH_2PO_4 (13.8g/L) and 95 mL of 100mmol Na_2HPO_4 (14.2g/L) were mixed. pH of the resulting solution was checked (if too low dibasic salt was added to reach pH = 8 and if too high monobasic salt was added to reach pH = 8). 21.7mg of acetylthiocholine iodide was dissolved in 10 mL of the 100 mmol/L of sodium phosphate monobasic buffer.

For each farmer, whole blood ChE activity was assayed in triplicate according to procedures of Ellman *et al.,* (1961). The analytical reproducibility (CV %) ranged from 1.28% to 1.94% in serially drawn samples. 0.5 mL of blood sample was thawed and diluted to 10 ml with a freshly prepared 100 mmol/L sodium phosphate monobasic buffer (pH = 8). 30 ml 100 mmol/L sodium phosphate buffer (pH = 8) and 1 ml DTNB were added to 1 mL of the diluted blood. The solution was mixed thoroughly and incubated at 30 °C for 10 minutes. The spectrophotometric zero was set with this solution (3 ml pipetted into a curvette), after which 133 µl of acetylthiocholine iodide solution was added. The rate of reaction was monitored in kinetic mode by a spectrophotometer at 412 nm for 6 min at 30 °C. The enzyme activity was expressed as µmoles of thiocholine hydrolysed min/mL of whole blood using an extinction coefficient of 13,600 cm/M.

A structured questionnaire in English was used to collect information from the farmers. The principal investigator translated the questionnaire into local and easily understandable languages, taking care to retain their original meaning. In some instances the principal investigator sought assistance from other farmers to translate the questionnaire into the local language. The questionnaire consisted of both open and closed questions. We asked for four groups of data: (1) personal, (2) lifestyle, i.e. hygiene, eating, smoking and drinking habits, (3) farm details and work history and (4) pesticide use practices and management. Additional questions were asked on pesticide exposure symptoms during the preceding week. Some of the questions included in the questionnaire are not relevant to the present paper. The questionnaires were administered at various locations, including farmers' homes, farms and school classrooms.

SPSS software (SPSS software, version 12.0.1 for Windows, SPSS Inc, Chicago, Illinois, USA) was used for all statistical analyses. First, we summarised responses from the questionnaire over the exposed and control groups with regard to their demographic characteristics. The independent t-test was used for bivariate comparisons between means. Categorical variables were compared by the χ^2-test. Multiple linear regression analysis was carried out to evaluate the contribution of demographic characteristics and exposure to ChE activities. For the comparison of symptoms according to exposure, we used the odds ratio for the prevalence as a measure of association. A level of probability below 0.05 was considered statistically significant for all analysis. Two-tailed p-values are given throughout. The results of the ChE activities were analysed for normal distribution (Kolmogorov-Smirnov Test) and homogeneity of variance (Levene's Test).

Results

Subjects who refused to allow their blood samples to be taken were not evaluated, so the overall participation rate among the exposed and control groups were about 60%. The final study population, therefore, consisted of 63 exposed subjects and 58 unexposed controls. The sample size is considered adequate for the desired level of confidence and precision.

The normal reference mean ChE activity adopted in this study is the one calculated from the control group. No study had been done in Ghana to establish a reference whole blood ChE activity. However, a study conducted in South Africa that established a reference range for serum ChE indicates a fairly similar reference range as adopted in this study (Rama and Jaga, 1992). Ideally, it is best to compare the farmers' post-exposure ChE level with their own baseline value. This was not possible in this study. The levels of ChE determined here for the controls, thus, served as baseline values for comparisons of the health effects of pesticide exposure.

Table 5.2 depicts the demographic characteristics of the farmers at Akumadan and controls at Tono. The mean age of the exposed participants (33.9 years) was slightly older than that of controls (32.3 years), but the difference was not significant (p = 0.678) (see Table 5.2). The mean duration of work as a farmer of the exposed participants was 17.1 years and the mean of the control group was 14.2 years. This difference was also not significant (p = 0.177). Exposed participants, on the average, weighed heavier than the control participants, though the difference was not statistically significant (p = 0.156). On the other hand, the control particiants were taller than the exposed participants and again thedifference was not satistically significant (p = 0.220). Measures reflecting lifestyle choices did not differ between the two groups except for drinking. Self-report of smoking habits of the two groups was not significantly different (p = 0.514). However, drinking habits differed

Table 5.2. Demographic characteristics and ChE activity (µmol min/mL of whole blood) of exposed and control participants

Characteristic	Community		Test of significance	p-value
	Exposed participants $N = 63$	Control participants $N = 58$		
Age (years)				
Mean (SD)	33.9 (8.51)	32.3 (9.80)	-0.418[*]	0.678
Gender, N (%)				
Male	33 (52)	38 (66)	2.15[†]	0.143
Female	30 (48)	20 (34)		
Duration of work as farmer (years)				
Mean (SD)	17.1 (5.30)	14.2 (6.10)	1.36[*]	0.177
Height (m)				
Mean (SD)	1.61 (0.130)	1.64 (0.093)	-1.23[*]	0.220
Weight (kg)				
Mean (SD)	63.8 (9.90)	61.2 (10.2)	1.43[*]	0.156
Smoking, N[‡] (%)				
Yes	7 (11.1)	3 (5.17)	0.672[†]	0.514
No	56 (88.9)	43 (74.1)		
Drinking, N[‡] (%)				
Yes	13 (20.6)	33 (56.9)		
No	49 (77.8)	13 (22.4)	27.84[†]	< 0.001
Mean ChE (SD)	3.59 (2.93)	7.27 (1.71)	-8.506[*]	< 0.001
% Falling below a level at 70% of normal reference mean [a]	73.0	13.8		

* Independent t-test. [†] χ^2-test. [‡] Deviation from N is due to missing answer to the specific question. [a] Normal reference mean: 7.27 µmol min/mL of whole blood

Table 5.3. Multiple linear regression model for predicting the relationship between ChE activity and demographic characteristics of exposed and control participants

Variable (unit)	Exposed (N = 63)			Control (N = 58)		
	β (SE)	Test of significance	p-value	β (SE)	Test of significance	p-value
Age (years)	-0.015 (0.046)	-0.318	0.751	0.003 (0.019)	0.158	0.875
Sex (female = 0, male = 1)	0.181 (1.19)	0.152	0.880	0.419 (0.741)	0.566	0.574
Weight (kg)	-0.006 (0.042)	-0.132	0.895	-0.001 (0.024)	-0.047	0.962
Height (m)	0.264 (3.11)	0.085	0.932	-0.994 (2.56)	-0.389	0.699
Constant	4.04 (5.70)	0.708	0.482	8.42 (4.26)	1.98	0.053
R^2	0.003			0.010		

Table 5.4. Prevalence (%) and odds ratio of self-reported symptoms attributable to pesticide exposure in exposed and control participants

Self-reported symptoms	Exposed (N = 62) N (%)	Control (N = 58) N (%)	OR	95% CI	p-value
Vomiting	2 (3.2)	5 (8.6)	0.348	0.065 to 1.87	0.218
Weakness	31 (50.0)	12 (20.7)	3.71	1.66 to 8.30	0.001
Headache	20 (32.3)	19 (32.8)	0.955	0.445 to 2.05	0.905
Itching	2 (3.2)	3 (5.2)	0.601	0.097 to 3.73	0.585
Stomach pain	5 (8.1)	4 (6.9)	1.16	0.297 to 4.56	0.828
No symptoms	2 (3.2)	15 (25.9)	0.094	0.020 to 0.432	0.002

OR, odds ratio; CI, confidence interval

significantly ($p < 0.001$) for the exposed and control groups.

The mean whole blood-ChE activity of farmers at Akumadan (3.59 μmol min/mL blood) was 50.6% less than that of farmers considered unexposed from Tono Irrigation Project (7.27 μmol min/mL blood), and the difference was highly significant ($p < 0.001$). Levels of cholinesterase were thus significantly lower in exposed farmers than in controls. The percentage of exposed vegetable farmers with a reduction of 30% in whole blood ChE activity was about 73% (Table 5.2). Although the groups did not differ significantly with respect to age, sex, height, and body weight, these variables are so well established as confounders of ChE activity in humans (Nielsen and Andersen, 2002), that multiple linear regression analysis was used to adjust for their influence. After adjustment for confounders, whole blood ChE was not influenced significantly for either exposed ($F = 0.043$, $df = 4, 56$, $p = 0.996$) or control group ($F = 0.128$, $df = 4, 52$, $p = 0.972$). None of the four confounders has significant influence on whole blood ChE (Table 5.3). Furthermore, work practices that encourage high exposures and, therefore, hypothesized to be associated with changes in ChE activity had no significant effect ($F = 0.328$, $df = 4,57$, $p = 0.858$) on ChE activity of exposed participants. Clearly, none of the four practices (method of pesticide application, direction of spraying of pesticide, kind of protective cover, and farmer re-entry period) was significantly associated with changed concentrations of ChE activity (Table 5.4).

Table 5.4 Multiple linear regression model for predicting the relationship between ChE activity and work practices that encourage high exposures and poisoning poten tials among exposed farmers ($N = 63$)

	β (SE)	Test of significance	p-value
Method of pesticide application	0.271 (0.724)	0.375	0.709
Direction of spraying of pesticide	0.197 (0.292)	0.675	0.503
Kind of protective cover	0.748 (1.06)	0.703	0.485
Farmer re-entry periods	0.269 (0.559)	0.481	0.632

Farmers recalled symptoms, attributable to pesticide exposure, they had experienced within the week preceding the interview. Self-reported symptoms of body weakness, headache, and stomach pain were higher in the exposed than in the control group. The difference was significant ($p < 0.05$) for body weakness alone. While self-reported symptoms of vomiting and body itching were lower in the exposed participants, they were not significant (Table 5.5). Exposed participants reported significantly ($p < 0.05$) more symptoms attributable to pesticide exposure than the controls.

Discussion

Our study demonstrated a statistically significant ($p < 0.001$) pattern of lower blood ChE activity in farmers at Akumadan compared to controls. The WHO biological index for individuals occupationally exposed to pesticides has been set to 30% of ChE depression (WHO, 1982). Accordingly, in the present study a ChE depression at or above 30% was selected to quantitatively estimate exposure to pesticides. Thus, any factor associated with a ChE depression greater than 30% could be attributed to pesticide exposure. In spite of this, approximately 73% of all exposed farmers had blood ChE activity levels at or below 70% of the normal reference mean. Acute overexposure and chronic moderate exposure to pesticides can result in an inhibition of ChE. However, symptoms depend more on the rate of fall in ChE activity than on the absolute level reached (Barnes, 1999).

The observation of lowered blood ChE activity in farmers at Akumadan is probably due to unsafe working habits promoted by self-desires to have rapid knockdown of pests and increase income. Akumadan lies in the agricultural zone of the Ashanti Region of Ghana. The community at Akumadan is noted for the cultivation of various vegetables: pepper, eggplants, onions, tomatoes, okra etc. and pesticide applications frequently occur. Both anecdotal evidence and available data of farmers' use of pesticides at Akumadan indicate that they not typically utilise recommended doses nor do they utilise the chemical industry's recommended practices for safe storage, handling, and application. There appear to be many reasons for this state of affairs (Ntow *et al.*, 2006). On the one hand, most farmers at Akumadan have little (64.9% comprising primary and middle) or no formal education (12.7%) (Ntow *et al.*, 2006). While their traditional farming knowledge is adequate for them to benefit from the trade, effective and safe chemical pest management requires knowledge that goes beyond traditional farming practices (Antle and Pingali, 1995). A wide and changing array of insecticides, herbicides, and other pesticides are available to farmers, but little extension is available to guide farmers in their use, and most have not received training for safe storage, handling, and application. Most farmers rely on the recommendations of chemical dealers or their own experience in deciding how to use pesticides. Majority (66.7%) of the farmers represented in this study used knapsack sprayers to apply toxic pesticides but none took effective self-protection measures when working in the field with these materials.

The pattern of lowered blood ChE activity in exposed participants compared to controls was supported by multiple regression analysis, which adjusted for age, sex, body weight, and height, strong contributors to ChE activity in humans (Nielsen and Andersen, 2002). The decreases in ChE activity were seen in self-reported symptoms attributable to pesticide exposure, particularly body weakness and headache (Table 5.5). About 97% of the population of vegetable farmers at Akumadan (exposed group) have reported symptoms attributable to pesticide exposure in the week preceding the survey. The symptom significantly ($p < 0.05$) reported by farmers is body weakness. This symptom is considered a common manifestation of cholinesterase inhibition (Quinones *et al.*, 1976; Yassin *et al.*, 2002).

In the questionnaire we had a strong focus on work practices and the use of personal protective equipment. Work practices involving use of personal protective clothing, method of application of pesticide, farmer re-entry intervals, and direction of spraying of pesticides were not associated with changed concentrations of ChE activity, although all four practices are hypothesized to be associated with changes in ChE activity. These practices are not governed by strict regulations concerning the training and education of farmers. As information from the questionnaire and previous work (Clarke *et al.*, 1997; Ntow *et al.*, 2006) revealed that farmers' handling and storage of chemical pesticides, personal hygiene and the proper use of personal protective equipment in Ghana is below a reasonable standard, it is unexpected that the biological exposure marker failed to demonstrate an equivalent result.

The interpretation of the relations in this study, however, is not unequivocal and is complicated by some methodological inadequacies in the study. A small sample size limits the significance of this study. However, it paves the way for a larger Ghanaian study with extended practical significance.

Other study limitations included the possible influence of bias. Most of the data were self reported, which could introduce a strong information (responder) bias, because respondents might have presented inaccurate information either in the hope of secondary gain or to avoid adverse outcomes-such as being investigated by authorities. Additionally, the lack of income for the control farmers who had not been farming for two months might have caused stress and related problems, possibly causing bias in the survey.

The criterion used for classifying study participants into exposed and a control group makes it difficult to reach significant differences, since they were all farmers with either direct or indirect exposure to pesticides. It would have been of most interest to monitor changes in blood ChE activity of the study population during the agricultural season in which each farmer could then serve as his control. This might weaken our study by making associations more difficult to detect, although it does not undermine the validity of any observed associations.

Policy implications

Our findings stress the need to review safety precautions in the use and application of pesticides in Ghana. The abuse, misuse, and the use of a wide range of pesticides, mostly of moderate to high toxicity in the country, implies high exposures and possible poisoning potentials and would suggest the need for more control and monitoring at national and local levels.

An immediate priority in Ghana is an urgent requirement for sustained, low cost, and well-targeted training interventions.

Acknowledgements

The authors thank Augustina Kumapley, Rafia Mamudu, and vegetable farmers in Ghana for their assistance and cooperation. This study was funded by the Dutch Government (through the UNESCO-IHE Institute for Water Education, Delft), International Water Management Institute, and the International Foundation for Science.

References

Antle JM, Pingali PL (1995) Pesticides, productivity, and farmer health: A Philippine case study. In: P.L. Pingali and P.A. Roger (eds.). Impact of pesticides on farmer health and the rice environment. Kluwer Academic Publishers, Massachusetts, pp 361-387.

Barnes JM (1999) Problems in monitoring overexposure among spray workers in fruit orchards chronically exposed to diluted organophosphate pesticides. Int Arch Occup Environ Health 72(Suppl 3):M68-M74.

Clarke E, Levy LS, Spurgeon A, Calvert LA (1997) The problems associated with pesticide use by irrigation workers in Ghana. Occup Med 47: 301-308.

Coye M, Barnett PG, Midttling JE (1986) Clinical confirmation of organophosphate poisoning of agricultural workers. Am. J. Ind. Med. 10:399-409.

Ellman GL, Courtney KD, Andres Jr V, Featherstone RM (1961) A new and rapid colorimetric determination of acetylcholinesterase activity. Biochemical Pharmacology 7:88-95.

Farahat TM, Abdelrasoul GM, Amr MM, Shebl MM, Farahat FM, Anger WK (2003). Neurobehavioural effects among workers occupationally exposed to organophosphorus pesticides. Occup. Environ. Med. 60:279-286. doi:10.1136/oem.60.4.279 (Online 14 November 2005).

Fenske RA, Lu C, Curl CL, Shirai JH, Kissel JC (2005) Biologic Monitoring to Characterise Organophosphorus Pesticide Exposure among Children and Workers: An Analysis of Recent Studies in Washington State. Environ Health Perspect 113(11):1651-1657.

He F (1993) Biological monitoring of occupational pesticides exposure. Int Arch Occup Environ Health 65:S69-S76.

Hogg RV, Tanis EA (1997) Probability and statistical inference, 5th ed. New Jersey: Prentice-Hall.

Nielsen JB, Andersen HR (2002) Cholinesterase Activity in Female Greenhouse Workers – Influence of Work Practices and Use of Oral Contraceptives. J Occup Health 44:234-239.

Ntow WJ (2001) Organochlorine Pesticides in Water, Sediment, Crops and Human fluids in a farming community in Ghana. Arch. Environ. Contam. Toxicol., 40(4):557-563.

Ntow WJ, Gijzen HJ, Kelderman P, Drechsel P (2006) Farmer perceptions and pesticide use practices in vegetable production in Ghana. Pest Manag Sci, 62:356-365.

Nurah GK (1999) A baseline study of vegetable production in Ghana. National Agricultural Research Project (NARP) Report, Accra.

Ohayo-Mitoko GJA, Kromhout H, Simwa JM, Boleij JSM, Heederik D (2000) Self reported symptoms and inhibition of acetylcholinesterase activity among Kenyan agricultural workers. Occup. Environ. Med. 57:195-200. doi:10.1136/oem.57.3.195 (Online 14 November 2005).

Quinones MA, Bogden JD, Louria DB, Nakah AE (1976) Depressed cholinesterase activities among farm workers in New Jersey. The Science of the Total Environment 6:155-159.

Rama DBK, Jaga K (1992) Pesticide exposure and cholinesterase levels among farm workers in the Republic of South Africa. The Science of the Total Environment 122:315-319.

Wadsworth Jr. HM (1990) Handbook of Statistical Methods for Engineers and Scientists. McGraw-Hill, Inc., New York

WHO Study Group (1982) Recommended health-based limits in occupational exposure to pesticides. Technical Report Series 677, World Health Organization, Geneva.

WHO Task Group (1986a) Environmental Health Criteria 63. Organophosphorus Insecticides: A General Introduction, World Health Organization, Geneva.

WHO Task Group (1986b) Environmental Health Criteria 64. Carbamate Pesticides: A General Introduction, World Health Organization, Geneva.

Yassin MM, Abu Mourad TA, Safi JM (2002) Knowledge, attitude, practice, and toxicity symptoms associated with pesticide use among farm workers in the Gaza Strip. Occup Environ Med 59:387-393.

Chapter 6

Accumulation of persistent organochlorine contaminants in milk and blood serum of vegetable farmers

Publication based on this chapter:

William J. Ntow, Laud M. Tagoe, Pay Drechsel, Peter Kelderman, and Huub J. Gijzen (2007). Accumulation of persistent organochlorine contaminants in milk and serum of farmers from Ghana. *Environmental Research* 106:17-26.

Accumulation of persistent organochlorine contaminants in milk and blood serum of vegetable farmers

Abstract

In the present study the concentrations of persistent organochlorine (OC) pesticides such as dichlorodiphenyltrichloroethane and its metabolites (DDTs), hexachlorocyclohexane isomers (HCHs), hexachlorobenzene (HCB) and dieldrin in pooled samples of human breast milk (n = 109), and serum (n = 115) from vegetable farmers in Ghana, during 2005, were determined. Gas chromatography with mass spectrometry was used to quantify residue levels on a lipid basis of the OCs. The pattern of OCs in human fluid showed that DDTs was consistently the prevalent OC in milk and blood. The level of DDTs, HCHs and dieldrin in the breast milk samples was found to correlate positively with age of the milk sample donors (r_s = 0.606, r_s = 0.770 and r_s = 0.540, respectively). When blood serum levels of the OCs were compared between male and female farmers, no pronounced relationship for HCHs and HCB ($p > 0.05$) was observed. However, DDTs and dieldrin residues were significantly higher ($p < 0.05$) in males than in females. There was association between breast milk and serum residues. When daily intakes of DDTs and HCHs to infants through human breast milk were estimated, some individual farmers (in the case of DDTs) and all farmers (in the case of HCHs) accumulated OCs in breast milk above the threshold (tolerable daily intake, TDI, guidelines proposed by Health Canada) for adverse effects, which may raise concern on children health.

Keywords: Contamination; environment; pesticides; pollutants; organochlorine

Introduction

Organochlorine (OC) compounds are a wide group of chemicals, many of which persist in the environment. Most of these OCs are agricultural pesticides or industrial compounds that also double as environmental pollutants. In response to their adverse effects seen in wildlife and humans, the production and use of these compounds were banned in industrialised countries during the 1970s (Jaraczewska et al., 2006), or subjected to restrictions in use in many others. However, they continue to be detected in both biological and environmental samples worldwide because of their persistent and bioaccumulative properties. There is much evidence to show that OCs interact with the endocrine system, resulting in numerous biologic effects that may affect the health of humans and animals (Muñoz-de-Toro et al., 2006).

The toxicological effects of OC contaminants create special problems for individuals who are exposed to them. Most exposure to these chemicals occur via occupation, dietary intake especially food of animal origin, water, ambient and indoor air, dust and soil (Cruz et al., 2003). Once they enter the biological system, these lipophilic compounds accumulate and even biomagnify their concentration along the food chain, especially in fatty food, and thus within the food chain bring on a high degree of contamination in high trophic organisms (Poon et al., 2005). Because humans occupy the top position in the trophic levels, they are obviously exposed to a higher level of these contaminants from aquatic and terrestrial food chains and become vulnerable to the toxic effects. Human beings, like other mammals, have lipid-rich tissue that efficiently retains and accumulates lipophilic contaminants (Travis et al., 1988); and they transfer much of the contaminant loads to offspring during nursing, resulting in trans-generational transfer of contaminants (Kanja et al., 1992; Waliszewski et al., 2001; Muñoz-de-Toro et al., 2006). The concentration of fat-soluble contaminants like the OCs is expected to be considerably higher in breast milk than in whole blood because the blood flow to the breast is much more rapid than is the rate of milk secretion (Weisenberg *et al.*, 1985).

Since 1970, body burden measurements for OCs in humans have been determined in serum, follicular fluid, human milk, and adipose tissue. Collection of adipose tissue involves invasive surgical removal of tissues. Therefore, a biological matrix, such as blood and breast milk or their components, that can be obtained with a less invasive procedure than adipose tissue is highly desirable (Pauwels *et al.*, 2000) and they have become common tissues for measuring OC levels in humans (Cerrillo *et al.*, 2005; Kunisue *et al.*, 2004a, 2004b; Minh *et al.*, 2004; Sudaryanto *et al.*, 2005; Sun *et al.*, 2005; Waliszewski *et al.*, 2001). However, serum is usually preferred over whole blood due to the minor complexity of the matrix and because it is a more homogenous material (Lino and Silveira, 2006). While several reports on organochlorine contaminants in human tissues (adipose tissue, milk or serum) have been published for other countries (Carreño *et al.*, 2006; Cerrilo *et al.*, 2005; Jaraczewska *et al.*, 2006; Kunisue *et al.*, 2004a, 2004b; Lino and da Silveira, 2006; Muñoz-de-Toro *et al.*, 2006; Pauwels *et al.*, 2000; Poon *et al.*, 2005; Sun *et al.*, 2005; Waliszewski *et al.*, 2001), there is insufficient documentation for Ghana, especially for recent years.

In this study, OC pesticides were measured in samples of human milk and blood serum from the vegetable farming population of Ghana. Our aim was to determine the body burden of OC pesticide residues in individuals who are placed in areas of intense agricultural activity, and for that, with higher environmental exposure and evaluate the health hazards. We intended to provide useful data on contamination levels in farmers in a region of Africa that has not been well studied. An additional source of interest could be to compare the possible differences between milk and serum from the same donors, milk concentrations and mothers' age, and between serum concentrations and gender. All the samples were analysed for the following residues: hexachlorocyclohexane (HCH) isomers (α, β, δ), dieldrin, hexachlorobenzene (HCB), 1,1,1-trichloro-2,2-bis(*p*-chlorophenyl) ethane (*p,p'*–DDT), 1,1-dichloro-2,2-bis(*p*-chlorophenyl) ethylene (*p,p'*–DDE), and 1,1-dichloro-2,2-bis(*p*-chlorophenyl) ethane (*p,p'*–DDD). These contaminants belong to the organochlorine class of pesticides, which are no longer used in many countries. However, despite the ban and restriction on their usage, in some cases, they are still used or they are present as persistent residues of previous uses.

Materials and methods

Study area
For practical reasons, the study was carried out in Ghana in Offinso District (Ashanti Region) and the Tono Irrigation Project in Kassena-Nankana District (Upper East Region) as shown in Fig. 4.1 (see chapter 4, between February 2005 and January 2006. These areas are characterised by intensive agricultural activities and thus are prone to higher incidences of environmental and dietary exposure to OC pesticides. More than 23 different active ingredients formulated as insecticides, herbicides and fungicides are reported (Ntow *et al.*, 2006) to be used in agriculture in the areas.

Offinso District has 126 settlements including five towns namely Abofour, Nkenkasu, Afrancho, Akumadan and New Offinso. The total population of the district is about 140,000. The population between the ages 15 and 64 (who are economically active) is 49% of the total population. The male to female ratio is

1:1.01. In the present study, farmers were selected from Nkenkasu, Afrancho and Akumadan. The major crops cultivated in the district are cassava, maize, plantain, vegetables, oil palm and cocoa. Kassena-Nankana District presides over the Tono Irrigation Project through ICOUR (Irrigation Company of the Upper Region), a government project started over a decade ago to promote the production of food crops (particularly vegetables) by small scale farmers within organised and managed irrigation schemes. The project, which uses the waters of River Tono (Fig. 4.1, chapter 4) which is dammed for the purpose, is divided amongst 3,000 farmers. The male to female ratio is 1:1.03. The cropping areas of the project are divided 50:50 between upland and lowland areas. Crops grown in upland plots include vegetables, millet, groundnut, sorghum and maize.

Cohort establishment
The target population was farmers, of either gender, 18 years or above, from the two agricultural areas working in vegetable farms. The farmers have applied pesticides between 1 and 26 years, and are still applying pesticides. The selection of the subjects was random. The estimated number of farmers using pesticides in Offinso District was 96% (Abraham Owusu, personal communication), and 97% in Kassena-Nankana District (personal communication with ICOUR). The sample size was determined in order to have 95% confidence limits of 5% maximum error of the estimate, when the prevalence is 96% for Offinso District, and 97% for Kassena-Nankana District (Hogg and Tanis, 1997; Wadsworth Jr., 1990). This leads to a requirement of about 60 and 45 farmers for Offinso and Kassena-Nankana Districts, respectively. For a no-response expectation (that is, subjects whose samples would be excluded from analytical measurements because of small volumes), the sample size was increased to 100 farmers for each district.

Ethics
All donors gave informed consent after receiving oral and/or written information on the project. The Ethics Review Committee of the Health Research Unit of the Ghana Health Service approved the study.

Sampling
OC residues were evaluated in blood serum from 115 subjects, 59 females and 56 males (ages of men and women were between 18 and 74 (mean = 35) and 18 and 53 (mean = 36), respectively). Blood was drawn with heparin-treated syringes and needles with the assistance of health workers and transferred to 5 mL disposable plastic tubes. The serum was separated by centrifugation (1500 rpm for 5 minutes) using the Beckman J2-21 Centrifuge and frozen at −4 °C in glass vials (pre-washed with *n*-hexane) until analysis.

Breast milk (n = 109; 45 females from which milk samples were taken also donated blood) was expressed manually into sterilized glass containers by lactating mothers. The women's ages were between 18 and 40 (mean = 28). All samples were stored at −4 °C until analysis. The samples were taken from mothers with normal and healthy babies who had reported at the clinic for postnatal observations.

Sample Analysis
All solvents and reagents used were pesticide-scan grade (Sigma; Munich, Germany; BDH; VWR International, UK; or Fluka; Munich, Germany). Supelclean Envi-Florisil SPE tubes (Supelco, Bellefonte, PA, USA) were used for sample cleanup processes. Anhydrous sodium sulphate was cleaned and activated at 450 °C

for 4 h. Reference analytical standards were supplied by Dr Ehrenstorfer, Augsburg, Germany.

The analysis of the OC compounds was performed according to method 2 of Lino *et al.* (1998). Briefly, the extraction of the OC compounds was made in 1 mL of serum with 2 x 5 mL of *n*-hexane-acetone (90 + 10), shaken for 1 min on vortex mixer and centrifuged at 1250 *g* for 5 min. The extract was transferred to a Florisil SPE cartridge previously added with 1 cm of sodium sulfate. Two different eluents were used: E_1 6 mL *n*-hexane and E_2 6 mL *n*-hexane-dichloromethane (5 + 1). The eluent was concentrated by a stream of nitrogen gas to 1 mL and then, quantification and confirmation of results were made using gas chromatography-electron capture detection (GC-ECD) and gas chromatography-mass spectrometer (GC-MS). Complete details of the gravimetric measurement of the lipid percentage of blood serum are described elsewhere (Pauwels *et al.*, 2000).

Samples were prepared following a previously reported methodology (Poon *et al.*, 2005). Briefly, about 2 g of human milk were weighed in a 100 mL reagent bottle and homogenized with 10 g anhydrous sodium sulfate with a stainless steel blender for 1 min. The sample was then extracted with 50 mL 1:1 acetone and hexane in a shaking incubator at 40 °C for at least 12 h. The extraction process was repeated. The extract was then concentrated to 5 mL using a rotary evaporator at 80°C. One-fifth of the concentrated extract was used for fat content determination by gravimetric method (Kunisue *et al.*, 2004a). The extract was transferred to a Florisil SPE cartridge previously added with 1 cm of sodium sulfate for cleanup. The column was washed four times with 8 mL *n*-hexane and the combined eluent was concentrated to 0.5 mL under reduced pressure. The final eluent was diluted to 1.5 mL with *n*-hexane.

Analysis of OC compounds was first performed on a Perkin Elmer AutoSystem GC with [63]Ni electron capture detector. A capillary column (methyl phenyl phase, 30 m * 0.32 mm I.D * 0.25 μm film thickness) was used with helium as carrier gas at a flow rate of 16 mL/min. Nitrogen make-up gas was used in the detector at 30 mL/min. Sample volumes of 1 μL were injected. Split injection of sample was undertaken at 250 °C. The column oven was programmed from an initial temperature of 100 °C and held at that temperature for 1 min. Programming rate was 10 °C min (100 to 150 °C); 5 °C min (150 to 250 °C); 30 °C min (250 to 300 °C) and held for 10 min at 300 °C. Detector temperature was 350 °C. Data were collected on a computer and reprocessed using Turbochrome Workstation. The identity of the OC compounds in serum and milk was confirmed by GC-MS (Agilent 6890 Series GC System with mass selective detector Agilent 5973N and fused capillary column (HP-5MS) packed with 5% phenyl methyl siloxane (30 m * 0.25 mm I.D and film thickness 0.25 μm). The operating temperatures of the GC were: injection port 250 °C (splitless, pressure 22.62 psi; purge flow 50 ml min; purge time 2.0 min; total flow 55.4 ml min). Column oven: initial 70 °C, held 2 min, programming rate 25 °C min (70 to 150 °C); 10 °C min[-1] (150 to 200 °C); 8 °C min (200 to 280 °C) and held 10 min at 280 °C. The carrier gas was nitrogen at 15 psi; detector make-up, 30 mL/min. The injection volume was 1 μL (Agilent 7683 Series injector).

Analyte recovery experiments were performed with sample matrices and blank. Human serum (1 mL) and breast milk (2 g) samples were spiked with 0.01, 0.05 and 0.1 ng/mL of each pesticide standard. The samples were extracted and analysed, in accordance with the previously noted procedures. Recovery of the different pesticides ranged between 81% and 93% with the variation coefficients not exceeding 25%. Three replicates of samples were used. During the sample extraction, blanks were regularly processed (one in ten). Duplicate samples of human serum and breast milk were taken from subjects and analysed. Three replicate dilutions of each sample were used in the analysis. For quality control of gas chromatographic conditions, a checkout procedure detailed elsewhere (Ntow *et al.*, Unpublished Report) was performed before sample analysis. The limits of quantification (LOQs), based on ten times the average standard deviation (SD) of the blank ranged between 0.001 and 0.01 ng/g. Because most of the OCs analysed by GC-MS had an LOQ at or below 0.01 ng/g, the reporting limit was chosen as 0.01 ng/g for these compounds.

Statistical Analyses

Statistical analyses were conducted using the statistical software SPSS (release 11.0; SPSS Inc, Chicago, Illinois, USA). Concentrations of OC pesticides were expressed as arithmetic means ± standard deviations. Descriptive statistics were used to characterise the contaminants in samples of serum and milk. The Mann-Whitney *U* test was employed to detect differences of OC concentrations in milk and serum (collected from the same individuals, n = 45), and the differences of OC concentrations by gender, male (n = 59) and female (n = 56). Spearman's rank correlation was used to measure the strength of the association between the mother's age and pesticides concentrations in their milk (n = 51). A *p* value $\leq$ 0.05 was considered to indicate statistical significance.

Results

Breast milk (n = 109) and serum (n = 115) samples were analysed for the targeted OCs. The arithmetic mean and standard error of the mean of detected compounds are given in Table 6.1. We considered HCHs as the sum of the three isomers (α, β, δ) and DDTs as the sum of the three compounds: *p,p'*-DDE, *p,p'*-DDD, *p,p'*-DDT. DDTs, HCHs, dieldrin and HCB were found in farmers' fluids. DDTs were detected in 88% and 75% of all the breast milk and serum samples analysed, respectively. While all isomers of DDT were detected in breast milk, DDD was not detected in any of the serum samples. Among DDT isomers, DDE presented the highest levels, 44.8 and 7.1 ng/g in breast milk and serum, respectively. The mean level (lipid weight) of DDTs was 78.3 ng/g in breast milk and 9.1 ng/g in blood serum (Table 6.1). Mean total DDT levels were higher than the mean concentration of sum HCHs in both matrices. HCHs were found in breast milk and serum samples with a mean level (lipid weight) of 46.4 ng/g and 7.3 ng/g, respectively, and were detected in over 50% of each of the sample matrix (either breast milk or serum). The β- isomer of HCH was found in both breast milk and serum. Higher concentrations of HCHs were found in breast milk samples than in serum samples. Contrary to DDTs and HCHs, higher concentrations of dieldrin and HCB were found in serum samples than in breast milk samples. However, the differences were not very pronounced. Dieldrin and HCB appeared in at least 60% of each sample matrix (either breast milk or serum) analysed. The mean level (lipid weight) of dieldrin was 122.8 ng/g and 127.0 ng/g in breast milk and serum samples, respectively. The mean level (lipid weight) of HCB was 4.9 ng/g and 5.3ng/g in breast milk and serum samples, respectively.

Table 6.1. Levels of Organochlorine (OC) pesticides in milk and serum from farmers in Ghana

Components	Milk (n = 109)		Serum (n = 115)	
	Mean (ng/g lipid weight) ± SE	Frequency of detection (% > LOQ)	Mean (ng/g lipid weight) ± SE	Frequency of detection (% > LOQ)
DDTs				
p,p'-DDE	44.8 ± 4.2	86	7.1 ± 1.2	90
p,p'-DDD	8.0 ± 1.0	80	< LOQ	-
p,p'-DDT	31.4 ± 4.5	76	0.5 ± 0.1	90
Sum DDTs	78.3 ± 7.0	88	9.1 ± 1.3	75
HCHs				
α-HCH	192.0 ± 40.4	29	< LOQ	-
β-HCH	14.0 ± 2.3	43	0.2 ± 0.1	90
δ-HCH	< LOQ	n.d	4.1 ± 0.5	90
Sum HCHs	46.4 ± 5.5	69	7.3 ± 0.7	54
Dieldrin	122.8 ± 24.8	69	127.0 ± 27.2	74
HCB	4.9 ± 0.3	65	5.3 ± 1.9	84

Discussion

Occurrence of Pesticides

OC compounds such as DDTs, HCHs, dieldrin and HCB were found in farmers' fluids. Some of the OCs were detectable in one fluid tissue and not in another. Except for δ-HCH, which was not detected in human milk, and p,p'-DDD and α - HCH which were not detected in blood serum, all OCs were detected in the majority (> 50%, except α-HCH (29%) and β-HCH (43%)) of the samples (milk and blood) examined. Ghanaian vegetable farmers have been widely exposed to these contaminants and for a period ranging between 1 and 26 years. Studies done by Ntow et al. (2006) confirm this. The pattern of OCs in human fluid showed that DDTs was consistently the prevalent OC in milk and blood, indicating that DDTs is the major environmental contaminant in the Ghanaian environment. In fact, other monitoring studies (Amoah et al., 2006; Ntow, 2001; 2005; Ntow et al., Unpublished Report) also found DDTs in the environment of Ghana. Among DDT metabolites, p,p'-DDE was the predominant compound suggesting a wide use and long-term accumulation of DDTs in humans from Ghana. Although technical DDT is in the Provisional list of Banned Pesticides in Ghana, the Act (Act 528) establishing this ban (which started operation in April 1999) is not fully implemented. This, therefore, leaves room for non-compliance by pesticide dealers. In fact, DDT is still used in some parts of Ghana to immobilise fish in rivers (Ntow, 2001). DDT has also been used for malaria control programs. The stock of obsolete DDTs may still be illegally distributed and used. Interestingly, the p,p'-DDT to p,p'-DDE ratio, which may indicate whether the exposure was in the distant (low ratio) or recent (high ratio) past (Carreño et al., 2006), was estimated at 0.7 (for breast milk) and 0.07 (for serum) in the present study. These ratios suggest the influence of the prohibition of DDT and the decease in exposure to these compounds over the years.

Residue levels and accumulation

Concentrations of OC pesticides are presented in Table 6.1. Literature reports suggest different accumulation of individual OCs in different biological matrices. The different modes of origin and the different lipid contents probably lead to selective accumulation. The lipids of blood serum consist of phospholipids and two simple lipids, fatty acid esters of glycerol and cholesterol and free cholesterol and those of breast milk are mainly the simple glycerides (Pauwels *et al.*, 2000). Among the OCs analysed, dieldrin was found at highest levels in all the samples (milk and blood) up to 1265.4 ng/g (data not shown), followed by other compounds in the order of DDTs > HCHs > HCB with mean concentrations 1-4 orders of magnitude less than those of dieldrin. The serum levels of dieldrin (127.0 ng/g lipid) and HCB (5.3 ng/g lipid) were higher than the levels in breast milk (dieldrin, 122.8 ng/g lipid; HCH, 4.9 ng/g lipid). The concentrations of the OCs varied substantially among the subjects, presumably due to different amounts of exposure and metabolism differences among the subjects. A previous study (Ntow, 2001) conducted in similar subjects in a similar environment had shown that the mean levels of HCB in blood and milk (fat) of 30 and 40 µg/kg, respectively, and respective mean levels of *p,p'*-DDE (380;490) were lower than the levels reported for the same compounds in similar matrices in various industrial countries (Kanja *et al.*, 1992; Skaare *et al.*, 1988; Weisenberg *et al.*, 1985).

Specific accumulation according to sex
Using the blood serum of farmers, we compared the levels of OCs in male and female donors. The results of the comparison are shown for both genders in Figure 6.1 for DDTs, HCHs, dieldrin and HCB. The graphs show that the DDTs ($p < 0.05$) and dieldrin ($p < 0.05$) residues were significantly higher in males than in females. For instance, for DDTs, male farmers had a mean concentration of 10.6 ng/g, whereas in females it was 7.1 ng/g. For dieldrin, the male farmers had a mean concentration of 134.0 ng/g, whereas the female farmers had 114.7 ng/g (data not shown). It is known that adult female excrete lipophilic contaminants such as OCs via lactation and thus reducing the body burden of such contaminants. Mean total HCH concentration was higher in females than in males; other studies also have found higher mean total HCH concentration levels in females (Lino and Silveira, 2006). Despite all this, and because we have found statistically significant differences only for two compounds (DDTs and dieldrin) between the genders (Figure 6.1), always higher in males than in females, we could not prove that organochlorine concentrations in human serum are gender-dependent.

Concentrations of OCs in association with age of mothers
Concentrations of OCs in human breast milk vary with factors such as the age and number of children of the mother (Kunisue *et al.*, 2004a). The present study examined the relationship between concentrations of OCs in human breast milk and the age of mothers (Figure 6.2). We performed the simple regression analysis between the mothers' age and OC concentrations. Apart from HCB, which did not show statistical correlation ($p = 0.067$, $r_s = 0.323$), we observed statistical correlation for DDTs ($p < 0.001$, $r_s = 0.606$), HCHs ($p < 0.001$, $r_s = 0.770$), and dieldrin ($p < 0.01$, $r_s = 0.540$) (Fig. 6.2). The correlation was positive for these contaminants. This signifies that OC levels in human breast milk tend to increase with increase of mother's age. Kunisue *et al.* (2004b) reached the same conclusion in their investigation on Hong Kong and Guangzhou.

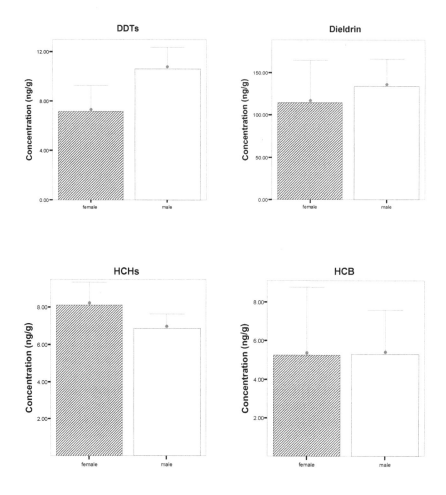

Figure 6.1. Comparison of blood organochlorine concentrations between male (n = 59) and
female (n = 56) farmers. For DDTs, $p < 0.05$; Dieldrin, $p < 0.05$; HCHs, $p = 0.376$;
and for HCB, $p = 0.910$. Nonparametric Mann-Whitney U-test was performed.

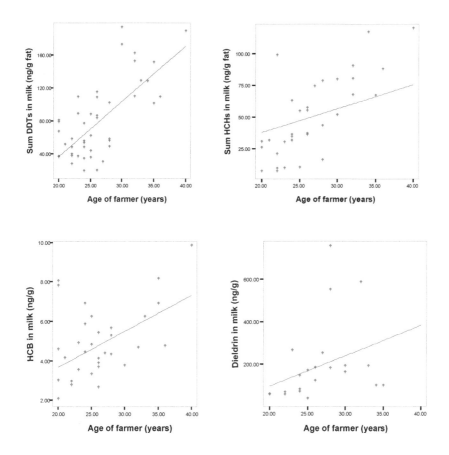

Figure 6.2. Relationships between mothers' age and breast milk concentrations of organochlorine contaminants in farmers (n = 51). Linear regression and Spearman rank correlation statistics for sum DDTs were $y = 6.72x + 98.09$, $r^2 = 0.50$, $r_s = 0.606$, $p < 0.001$; for sum HCHs, $y = 4.87x + 81.23$, $r^2 = 0.60$, $r_s = 0.770$, $p < 0.001$; for HCB, $y = 0.15x + 0.93$, $r^2 = 0.19$, $r_s = 0.323$, $p = 0.067$; and for Dieldrin, $y = 14.27x + 188.24$, $r^2 = 0.11$, $r_s = 0.540$, $p < 0.01$.

Relationship of OC pesticides in human milk and blood serum

A total of 45 pairs of samples (human milk and blood) were used to study their relationship with OCs contamination. Figure 6.3 depicts the correlation between the milk and blood samples for OC pesticide concentrations. Except dieldrin, for which there was no significant correlation ($p = 0.101$) in levels between breast milk and blood serum, DDTs, HCHs and HCB correlated ($p < 0.05$) for breast milk and blood serum. High levels of coherence of the persistent contaminants are in line with other studies (Kanja *et al.*, 1992; Pauwels *et al.*, 2000; Skaare *et al.*, 1988; Waliszewski *et al.*, 2001). It may be concluded that residues of OC pesticides in both human milk and blood serum could be used as indicators to show human accumulative exposure of these pollutants.

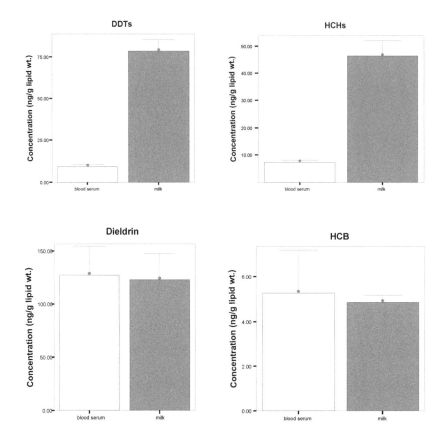

Figure 6.3. Comparison of organochlorine concentrations in human breast milk and blood (n = 45). Bars represent standard error of the mean. The p values were as follows: DDTs, $p < 0.05$; Dieldrin, $p = 0.101$; HCHs, $p < 0.05$; and HCB, $p < 0.05$. Nonparametric Mann-Whitney U-test was performed.

Risk assessment for infants

Estimated daily intake (EDI) of OCs by infants was calculated based on the assumption that the average milk consumption of a 5 kg infant is 700 g/day (Minh et al., 2004). The mean values of daily intake of organochlorines were estimated by using Eq. (6.1).

$$EDI = \frac{C_{milk} \times 700_g \times C_{lipid}}{5_{kg}} \tag{6.1}$$

Where EDI is the estimated daily intake (µg/kg body wt/day); C_{milk}: concentration of the chemical in milk ((µg/g lipid wt); C_{lipid}: lipid content in milk (%).

The estimated daily intakes are given in Table 6.2 and individual intakes are shown in Figure 6.4. It was recognized that although intake of DDTs by most infants is below the guideline (Tolerable Daily Intake, TDI) proposed by Health Canada (Oostdam *et al.*, 1999) in average, intake by some individuals is close to or exceeds this guideline. This fact may raise greater concerns on infant health because children are highly susceptible to effects from environmental contaminants.

Table 6.2. Estimated daily intake (µg/kg body wt/day) of OCs by infants in Ghana

		DDTs	Dieldrin	HCHs	HCB
Ghanaian infants	mean	12.4	25.1	7.1	0.8
	range	0.6-58.2	1.0-157.2	0.5-29.6	0.04-2.4
Tolerable Daily Intake (TDI)s[a]		20	-	0.3	0.3

[a] Cited from Oostdam *et al.*, 1999

There is evidence that *p,p'*-DDE is an androgen antagonist (Keice *et al.*, 1995). This compound is generally abundant among DDT metabolites in human and wildlife as evident in this study (Table 6.1). Furthermore, there is evidence that β-HCH is also an environmental estrogen (Willett *et al.*, 1998), and this compound is generally abundant among HCH isomers in human and wildlife as observed in this study (Table 6.1). These observations imply that abundance of these contaminants in human breast milk may adversely affect reproductive systems of Ghanaian children. To our knowledge, our study is the first to elucidate the contamination status of HCHs in human breast milk from Ghana.

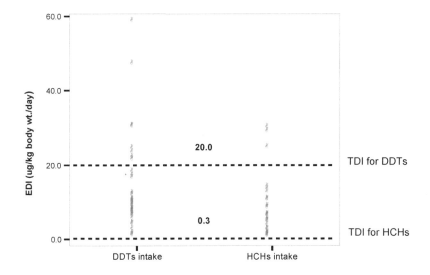

Figure 6.4. Estimated daily intake (DI) of DDTs and HCHs by infants in Ghana in comparison with the Tolerable Daily Intake (TDI) proposed by Health Canada (Oostdam *et al.*, 1999).

If only the risk of OCs is considered, breast milk containing high levels of such contaminants should not be fed to infants. Feeding with milk formula not contaminated by OCs instead of human breast milk is one of the measures for protecting infants from OCs risk. In human breast milk, however, not only general nutrients but also essential components for infant's growth and development such as secretory IgA, oligosaccharides, lactoferrin and lysozyme, which can increase their resistance to common infections, are present. In addition, long-chain polyunsaturated fatty acids in human breast milk are indispensable for brain development of infants and it is reported that breast-feeding is associated with significantly higher scores for cognitive development than formula feeding (Kunisue *et al.*, 2004b). So, avoiding breast-feeding may affect the growth and development of infants. Considering all these factors, breast-feeding is essential for the infants, but it is necessary to reduce the levels of DDTs and HCHs in human breast milk in Ghana. It is necessary to elucidate whether recent input and illegal use of DDTs and HCHs are present and to continuously investigate temporal trends of DDTs and HCHs pollution in Ghana to remedy the situation. However, in the viewpoint of environmental toxicology, elevated DDTs concentrations in Ghanaian human breast milk perhaps underlines higher risk to both mother and infant health and deserve stricter regulation to phase out completely the use of DDTs.

In summary, the present study is the most recent and comprehensive studies on OC pesticides contamination in human breast milk and blood serum in farmers from Ghana. Our results reveal the presence of persistent, bio-accumulative, and toxic DDTs, HCHs, dieldrin, and HCB in human fluids and at levels that raise public health concerns. Although the majority of epidemiological studies, have not confirmed these chemical compounds as likely causes of diseases like cancer, the fact that almost all samples tested to date have shown detectable levels of residues of pesticides (especially *p,p'*-DDE) provides ample reason for concern about many possible health effects of these compounds. Hence, further monitoring studies are needed in an attempt to eliminate or reduce sources of contamination, and providing information for epidemiological studies that establish a relation between levels of contamination and predominance of certain diseases like cancer.

Acknowledgements

The authors thank Augustina Kumapley, Rafia Mamudu, Sarah Kafui, Ato (Akumadan) and vegetable farmers in Ghana for their assistance and cooperation. Sampson Abu and Bernice Worlanyo Ntow are also acknowledged for their technical and secretarial assistance, respectively. This study was funded by the Dutch Government (through the UNESCO-IHE Institute for Water Education, Delft), International Water Management Institute, and the International Foundation for Science. The Kinneret Limnological Laboratory, Migdal, Israel, is acknowledged for technical assistance in the use of GC/MS.

References

Amoah, P., Drechsel, P., Abaidoo, R.C., Ntow, W.J., 2006. Pesticide and Pathogen Contamination of vegetables in Ghana's Urban Markets. Arch. Environ. Contam. Toxicol. 50,1-6.

Carreño, J., Rivas, A., Granada, A., Lopez-Espinosa, M.J., Mariscal, M., Olea, N., Olea-Serrano, F., 2006. Exposure of young men to organochlorine pesticides in Southern Spain. Environmental Research doi: 10.1016/j.envres.2006.06.007.

Cerrilo, I., Granada, A., López-Espinosa, M.-J., Olmos, B., Jiménez, M., Caño, A., Olea, N., Olea-Serrano, M.F., 2005. Endosulfan and its metabolites in fertile women, placenta, cord blood, and human milk. Environ. Res. 98, 233-239.

Cruz, S., Lino, C., Silveira, M.I., 2003. Evaluation of organochlorine pesticide residues in human serum from an urban and two rural populations in Portugal. Sci. Total Environ. 317, 23-35.

Hogg, R.V., Tanis, E.A., 1997. Probability and statistical inference, 5th ed. New Jersey: rentice-Hall.

Jaraczewska, K., Lulek, J., Covaci, A., Voorspoels, S., Kaluba-Skotarczak, A., Drews, K., Schepens, P., 2006. Distribution of polychlorinated biphenyls, organochlorine pesticides and polybrominated diphenyl ethers in human umbilical cord serum, maternal serum and milk from Wielkopolska region, Poland. Sci.Total Environ. doi:10.1016/j.scitotenv.2006.03.030.

Kanja, L.W., Skaare, J.U., Ojwang, S.B.O., Maitai, C.K., 1992. A comparison of organochlorine pesticide residues in maternal adipose tissue, maternal blood, cord blood, and human milk from mother/infant pairs. Arch. Environ. Contam. Toxicol. 22, 21-24

Keice, W.R., Stone, C.R., Laws, S.C., Gray, L.E., Kempainen, I.A., Wilson, E.M., 1995. Persistent DDT metabolite p,p'-DDE is a potent androgen receptor antagonist. Nature 375, 581-585.

Kunisue, T., Someya, M., Monirith, I., Watanabe, M., Tana, T.S., Tanabe, S., 2004a. Occurrence of PCBs, Organochlorine Insecticides, $tris$(4-Chlorophenyl)methane, and $tris$(4-Chlorophenyl)methanol in Human Breast Milk Collected from Cambodia. Arch. Environ. Contam. Toxicol. 46, 405-412.

Kunisue, T., Someya, M., Kayama, F., Jin, Y., Tanabe, S., 2004b. Persistent organochlorines in human breast milk collected from primiparae in Dalian and Shenyang, China. Environ. Pollut. 131, 381-392.

Lino, C.M., Azzolini, C.B.F., Nunes, D.S.V., Silva, J.M.R., Silveira, M.I.N., 1998. Methods for the determination of organochlorine pesticide residues in human serum. J. Chromatogr. Sec. B 716, 147-152.

Lino, C.M., da Silveira, M.I.N., 2006. Evaluation of organochlorine pesticides in serum from students in Coimbra, Portugal: 1997-2001. Environ. Res. 102, 339-351.

Minh, N.H., Someya, M., Minh, T.B., Kunisue, T., Iwata, H., Watanabe, M., Tanabe, S., Viet, P.H., Tuyen, B.C., 2004. Persistent organochlorine residues in human breast milk from Hanoi and Hochiminh city, Vietnam: contamination, accumulation kinetics and risk assessment for infants. Environ. Pollut. 129, 431-441.

Muñoz-de-Toro, M., Beldoménico, H.R., Garcia, S.R., Stoker, C., De Jesús, J.J., Beldoménico, P.M., Ramos, J.G., Luque, E.H., 2006. Organochlorine levels in adipose tissue of women from a littoral region of Argentina. Environmental Research 102, 107-112.

Ntow, W.J., 2001. Organochlorine pesticides in water, sediment, crops and human fluids in a farming community in Ghana. Arch. Environ. Contam. Toxicol. 40, 557-563.

Ntow, W.J., 2005. Pesticide Residues in Volta Lake, Ghana. Lakes & Reservoirs: Research and Management 10 (4), 243-248.

Ntow, W.J., Gijzen, H.J., Kelderman, P., Drechsel, P., 2006. Farmer perceptions and pesticide use practices in vegetable production in Ghana. Pest. Manag. Sci. 62 (4), 356-365.

Oostdam, J.V., Gilman, A., Dewailly, E., Usher, P., Wheatley, B., Kuhnlein, H., 1999. Human health implications of environmental contaminants in Arctic Canada: a review. Sci. Total Environ. 230, 1-82.

Pauwels, A., Covaci, A., Weyler, J., Delbeke, L., Dhont, M., De Sutter, P., D'Hooghe, T., Schepens, P.J.C., 2000. Comparison of Persistent Organic Pollutant Residues in Serum and Adipose Tissue in a Female Population in Belgium, 1996-1998. Arch. Environ. Contam. Toxicol. 39, 265-270.

Poon, B.H.T., Leung, C.K.M., Wong, C.K.C., Wong, M.H., 2005. Polychlorinated Biphenyls and Organochlorine Pesticides in Human Adipose Tissue and Breast Milk Collected in Hong Kong. Arch. Environ. Contam. Toxicol. 49, 274-282.

Skaare, J.U., Tuveng, J.M., Sande, H.A., 1988. Organochlorine pesticides and polychlorinated biphenyls in maternal adipose tissue, blood, milk and cord blood from mothers and their infants living in Norway. Arch. Environ. Contam. Toxicol. 17, 55-63.

Sudaryanto, A., Kunisue, T., Kajiwara, N., Iwata, H., Adibroto, T.A., Hartono, P., Tanabe, S., 2005. Specific accumulation of organochlorines in human breast milk from Indonesia: Levels, distribution, accumulation kinetics and infant health risk. Environ. Pollut. (doi:10.1016/j.envpol.2005.04.028).

Sun, S.-J., Zhao, J.-H., Koga, M., Ma, Y.-X., Liu, D.-W., Nakamura, M., Liu, H.-J., Horiguchi, H., Clark, G.C., Kayama, F., 2005. Persistent organic pollutants in human milk in women from urban and rural areas in northern China. Environ. Res. 99, 285-293.

Travis, C.C., Hattermer-Frey, H.A., Arma, A.D., 1988. Relationship between dietary intake of organic chemicals and their concentrations in human adipose tissue and breast milk. Arch. Environ. Contam. Toxicol. 17, 473-478.

Wadsworth Jr., H.M., 1990. Handbook of Statistical Methods for Engineers and Scientists. McGraw-Hill, Inc., New York

Waliszewski, S.M., Aguirre, A.A., Infanzon, R.M., Silva, C.S., Siliceo, J., 2001. Organochlorine Pesticide Levels in Maternal Adipose Tissue, Maternal Blood Serum, Unbilical Blood Serum, and Milk from Inhabitants of Veracruz, Mexico. Arch. Environ. Contam. Toxicol. 40, 432-438.

Weisenberg, E., Arad, I., Grauer, F., Sahm, Z., 1985. Polychlorinated Biphenyls and Organochlorine Insecticides in Human Milk in Israel. Arch. Environ. Contam. Toxicol. 14, 517-521.

Willett, K.L., Ulrich, E.M., Hites, R.A., 1998. Differential toxicity environmental fates of hexachlorocyclohexane isomers. Environ. Sci. Technol. 32, 2197-2207.

Chapter 7

Dietary exposure to pesticides from vegetables among adult farmers at Akumadan

Submission to a scientific journal based on this chapter:

William J. Ntow, Benjamin O. Botwe, Peter Kelderman, Pay Drechsel, and Huub J. Gijzen. Dietary exposure to pesticides from vegetables among adults in the farming community of Akumadan, Ghana. *Arch. Environ. Contam. Toxicol.* Submitted on March 15, 2007.

Dietary exposure to pesticides from vegetables among adult farmers at Akumadan

Abstract

The present study has determined the residues concentrations of pesticides in vegetables, and assessed the health risk due to the daily consumption of contaminated vegetables for adults in the prominent farming community of Akumadan. Fifteen pooled samples, belonging to five vegetable types (tomato, cabbage, pepper, onion, and eggplants) purchased from local markets across Ghana were analysed for DDTs, endosulfan, HCHs, methoxychlor, dimethoate, chlorpyrifos, dieldrin, HCB, heptachlor epoxide, and lambda-cyhalothrin. The concentrations of the pesticides ranged from 0.01 to 46.9 µg/kg wet weight. Dietary data were collected from 130 residents. By combining the dietary and contaminant data, dietary exposure to the identified pesticides was calculated. Dietary exposure to pesticide residues at Akumadan is low and there seems to be no associated health risk. However, the results of persistent contaminants are of particular health concern because of this persistence. The estimated exposure distributions were shown to be insensitive to valuation of the non-detect residue samples.

Keywords: contamination; environment; health risk assessment; pesticide residues; vegetables

Introduction

Pesticides applied to food crops in the field can leave potentially harmful residues. Organochlorine pesticides in particular can persist in foodstuffs for a considerable period. If crops are sprayed shortly prior to harvest without an appropriate waiting period, even organophosphate residues can persist up until the food is in the hands of the consumer (Bull, 1992). In Ghana, the increase in urban population and food demand has catalysed the use of chemical pesticides for food production (Amoah *et al.*, 2006).

Vegetables are grown extensively in Ghana, and constitute a large portion of the diet of the average Ghanaian. Vegetables are essential for a healthy and balanced diet, as well as adding variety, interest and flavour to the menu. However, vegetables also attract a wide range of pests and diseases, and can require intensive pest management. About 87% of the farmers who grow vegetables in Ghana use pesticides (Dinham, 2003). Many of these farmers spray the same wide range of pesticides on all vegetables and ignore pre-harvest intervals (Ntow *et al.*, 2006). Sometimes farmers spray pesticides one day before harvest to sell 'good-looking' vegetables. This practice, in particular, exposes consumers to pesticides. Though it is sometimes thought that residues are destroyed if food is properly washed and cooked, this is not always the case. Washing and cooking may reduce pesticide residues in food; boiling may remove only 35-60% of organophosphate residues and 20-25% of organochlorines (Bull, 1992). Residues above tolerance limits do occur in cooked food. Consumption of contaminated food is an important route of human exposure to pesticide residues, and may pose a public health risk (MacIntosh *et al.*, 1996).

This study, the first systematic assessment of possible health risk due to consumption of contaminated primary foods in Ghana, is part of a wider programme, which also contains a detailed questionnaire relating to practices of use of pesticides (Ntow *et al.* 2006). Specifically, the objectives of the study were:

- To measure pesticide residues concentrations in common vegetable items collected from local designated markets across Ghana
- To derive average daily exposure to the pesticides for a prominent vegetable growing community in Ghana based on the local diet consumption of vegetables
- To assess the potential health risks to human consumers based on this data.

The study was undertaken at the same prominent vegetable farming community in Ghana (Akumadan, see Figure 4.1, chapter 4) where toxicological investigation of human breast milk and blood serum had been carried out (see chapter 6). The last decade witnessed increasing concern for potential health effects for pesticide intake on the people of Akumadan, and this study was conducted at the request of the community members in the area.

Materials and Methods

Method of estimation of vegetable contamination exposures
The analysis (Dougherty *et al.*, 2000) used for this study estimates exposures to identified pesticides through consumption of selected vegetables. The analysis includes: (1) creating a vegetable consumption database that provides information on consumption patterns by the population, (2) creating a vegetable contaminant database by obtaining and compiling contaminant data, (3) combining contaminant and consumption information to estimate exposures from individual vegetable types, (4) estimating total vegetable dietary exposures by summing across all vegetable types, and (5) comparing exposures to benchmark concentrations to determine potential public health impacts. Figure 7.1 provides an overview of these steps.

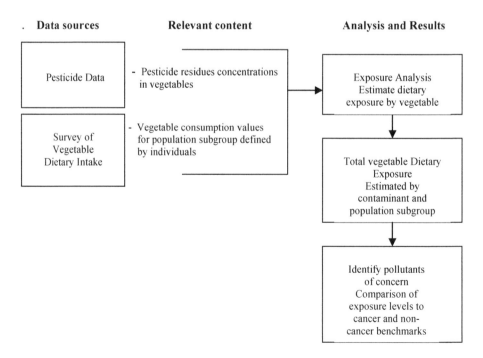

Figure 7.1. Overview of analytical methods for vegetable contamination exposures.

Sampling of vegetables

Identical vegetable items were purchased from local designated markets across all the 10 regions (Figure 7.2) in Ghana in 2005. The vegetable items, which were tomato, pepper, onion, cabbage, and eggplants, were considered representative of the diet of the Ghanaian population according to a survey in 2005 (Ntow *et al.*, 2006). At each market, vegetable samples were collected from three randomly selected sellers. Randomisation of sellers was accomplished by identifying all sellers of each vegetable item present on the day of sampling and randomly selecting three of them. Three composite samples of identical vegetable items-each containing 3 bulbs of tomato, 3 cabbages, 10 pepper fruits, 3 bulbs of onion, and 3 eggplants-were drawn from each market. Samples are considered representative of all commercially available produce, because each local designated market is the sole official fresh produce outlet. The samples (flesh) were given a cold-water wash with a soft brush to remove adhering soil particles, wrapped in aluminium foil according to type, packed in polythene bags, and stored in a freezer within 2-3 h of collection. Frozen samples were packed in insulated containers with dry ice and transported to the CSIR Water Research Institute Laboratory in Accra (1-2 days travel time) where they arrived still frozen. Samples were kept in a freezer at –4 °C until required for extraction, which was carried out within 24 h of arrival at the laboratory in Accra.

Sample preparation and extraction

Analytical standards of lambda-cyhalothrin (97.5%), DDTs (97.0-99.5%), dimethoate (98.0%), endosulfan (98.0-99.5%), HCHs (97.7-98.5%), chlorpyrifos (98.4%), dieldrin (98.3%), HCB (99.0%), heptachlor epoxide (99.0%), and methoxychlor (95%) were supplied by Dr Ehrenstorfer, Augsburg, Germany. All organic solvents used were of GC grade (Sigma, München, Germany or BDH, VWR International, UK).

In preliminary studies, the various sample types (homogenised vegetable samples placed in 250 mL standard joint borosilicate bottles) were spiked with appropriate volumes of previously prepared stock solutions of pesticide compounds (Ntow *et al.*, Unpublished). Each fortification level was prepared in three replicates. The bottles were capped, manually shaken to ensure thorough mixing, and stored in a deep freezer at – 4 °C for 24 h to simulate sample storage conditions. The average recoveries for all compounds varied from 83-91%. Triplicate analyses also gave a standard error of about 10%. Residue data were not corrected for efficiency of recovery. The limit of quantification (LOQ) was set to 0.01 µg/kg fresh weight for all compounds (calculated from real samples as being 10 times the signal to noise ratio). During sample extraction, blanks were regularly processed (one in ten).

The frozen samples were thawed and each cut into four equal segments (quartered). Opposite segments were discarded in order to reduce the bulk of the material needing to be processed. Following this step, the vegetables were cut into small pieces. For the analysis, samples of identical vegetable item were pooled as one composite. From this composite, three replicate samples were prepared for extraction and subsequent instrumental analysis. This design yielded three residue values for each vegetable item. Sample preparation and extraction followed the procedures described in Ferrer *et al.* (2005). About (10 g) of each vegetable was weighed into a porcelain mortar, and ground with 50 g of anhydrous sodium sulphate. The powdered sample was extracted in ethyl acetate. The extract was rotary evaporated at 40 °C, and the residue re-dissolved in hexane. Sample clean up followed the procedure of Hsu *et al.* (1991).

Instrumental analysis
The analysis of vegetable samples for pesticide compounds was performed on a GC-MS (Agilent 6890 Series GC System with mass selective detector Agilent 5973N and fused capillary column (HP-5MS) packed with 5% phenyl methyl siloxane (30 m x 0.25 mm I.D x 0.25 μm). The operating temperatures were: injection port 250 °C (splitless, pressure 22.62 psi; purge flow 50 mL/min; purge time 2.0 min; total flow 55.4 mL/min), oven 70 °C for 2 min, raised to 150 °C (25 °C/min), 200 °C (10 °C/min), then at 8 °C/min to 280 °C (held 10 min). The carrier gas was nitrogen at 15 psi; The injection volume was 1 μL (Agilent 7683 Series injector).

Ethics
All subjects who participated in the study gave informed consent after receiving oral and/or written information on the project. The Ethics Review Committee of the Health Research Unit of the Ghana Health Service had approved the study.

Dietary survey
In February 2005, as part of an ongoing research project on the use and fate of pesticides in vegetable-based agro-ecosystems in Ghana, a questionnaire-based dietary survey was conducted at Akumadan by randomly selecting and surveying 130 healthy adults from the farming population. Dietary data were collected during a detailed face-to-face interview, the details of which are provided in Ntow *et al.* (2006). We adopted the dietary history approach in our data collection (Kroes *et al.*, 2002). Subsequent to previous information about respondents' eating pattern over a week and the previous day, we completed a checklist of vegetables consumed by the subject over a "typical" week. Then, the quantity of identified vegetables consumed by each subject over a "typical" week was estimated by using vegetable models.

The vegetable consumption data derived from the survey of Ntow *et al.* (2006) are combined with this study's laboratory analysis results and mean body weight of subjects to calculate estimates of average daily intake of six tested pesticides.

Risk assessment
The total daily dietary exposure, E_t (mg/kg body weight) that results from eating a combination of contaminated vegetables was calculated as follows (USEPA, 1989):

$$E_t = \sum_{i=1}^{n} (C_f)_i (L)_i \qquad (7.1)$$

where C_f = the concentration (mg/kg) of the contaminant in the vegetable, L = the amount (kg/day/kg body weight) of contaminated vegetable consumed, and I = the number of different vegetable types consumed. For each contaminant, the average daily dietary exposure, E (mg/kg body weight), level for the population was calculated by:

$$E = C_f \times L \qquad (7.2)$$

Vegetable consumption was expressed as daily consumption divided by body weight.

To account for residues concentrations that were below the limit of quantification, we used two alternative assumptions. We first assumed that actual concentrations for non-detect (ND) samples were equal to the limit of quantification. However, this would overestimate the concentrations if there would be no contamination of the sampled vegetables. Therefore, we also calculated exposures with the alternate assumption that actual concentrations for all non-detect samples were equal to zero. These two assumptions provide a range of exposure estimates for each contaminant with non-detects.

To assess potential public health risks, exposure concentrations were compared to the minimal risk levels (MRLs) for the individual pesticide components. These were derived from the United States Agency for Toxic Substances and Disease Registry (ATSDR) in a manner similar to the way in which the reference dose (RfD) and reference concentration are determined by the United States Environmental Protection Agency (USEPA) (Jiang *et al.*, 2006). The oral MRLs are for: total DDTs 0.0005 (listed as p,p'-DDT); Endosulfan 0.002; HCHs 0.00001 (listed as γ-HCH); cyhalothrin 0.01; and methoxychlor 0.005 mg/kg/day (ATSDR, http://www.atsdr.cdc.gov/mrls.html, accessed February 26, 2007). There is no MRL or RfD available from ATSDR or USEPA for dimethoate. Here the EXTOXNET RfD value of 0.0002 mg/kg/day was applied (http://extoxnet.orst.edu/pips/ghindex.html, accessed February 27, 2007). The non-cancer health risks from the consumption of vegetables by Akumadan residents were assessed by estimating the relevant hazard ratios (HRs) (Eq. (7.3)). These hazard ratios were calculated by dividing the average daily exposure by the MRL. A HR > 1 indicates that the average exposure level exceeded the benchmark concentration (Dougherty *et al.*, 2000):

Hazard ratio (HR) = Average daily exposure/Minimal risk level (7.3)

Results

The results for pesticide residues that were detected and quantified in the five types of vegetable crops are shown in Table 7.1. Because of low concentrations of pesticides residues detected in vegetable samples, these results are expressed as micrograms per kilogram fresh weight, rather than as milligrams per kilogram.

The highest concentrations of pesticide residues were found in tomato, cabbage and onion (Table 7.1). These items also had the largest spectrum of contamination. With the exception of eggplants, all the vegetable items showed quantifiable amounts of HCHs (α, β, and γ). *p,p*'-DDE was almost ubiquitous, whereas other substances (e.g., *p,p*'-DDT, lambda cyhalothrin, and endosulfan (α, β, and sulphate) were less generally present in vegetables. Relatively high residue concentrations of methoxychlor were found in all vegetables except in cabbage. Mean concentrations of methoxychlor varied markedly among the vegetables, and were highest in onion (46.9 µg/kg), and lowest in cabbage (0.85 µg/kg). All samples were below quantification limits for chlorpyrifos, dieldrin, heptachlor epoxide, and HCB.

Among the 130 people (mean weight 67.2 ± 2.19 kg) interviewed, the mean age was 43 ± 11 years (range: 22-75), and more than half (59.2%) were men. According to the dietary survey, a healthy adult ate 565 ± 108 g (fresh weight) vegetables each day in a "typical" week.

Table 7.1. Concentrations (mean ± SE, μg/kg fresh weight) of pesticide residues in
vegetables from Ghanaian markets

Pesticide component	Tomato	Cabbage	Pepper	Onion	Eggplant
α-Endosulfan	0.06 ± 0.00	0.74 ± 0.00	< 0.01	0.55 ± 0.00	< 0.01
β-Endosulfan	0.02 ± 0.00	0.64 ± 0.00	< 0.01	0.37 ± 0.00	< 0.01
Endosulfan sulfate	0.13 ± 0.00	1.66 ± 0.01	< 0.01	0.93 ± 0.00	< 0.01
p,p'-DDT	0.01 ± 0.00	0.19 ± 0.01	< 0.01	< 0.01	< 0.01
p,p'-DDE	0.02 ± 0.00	0.26 ± 0.01	< 0.01	0.02 ± 0.00	0.01 ± 0.00
α-HCH	0.05 ± 0.00	0.14 ± 0.00	0.10 ± 0.00	0.03 ± 0.00	< 0.01
β-HCH	0.05 ± 0.00	0.15 ± 0.01	0.10 ± 0.00	0.02 ± 0.00	< 0.01
γ-HCH	0.12 ± 0.00	0.32 ± 0.00	0.15 ± 0.01	0.04 ± 0.00	< 0.01
Methoxychlor	28.2 ± 0.38	0.85 ± 0.02	7.43 ± 0.04	46.9 ± 0.34	1.00 ± 0.03
Dimethoate	0.01 ± 0.00	0.43 ± 0.01	< 0.01	1.76 ± 0.01	0.22 ± 0.00
Lambda-cyhalothrin	0.05 ± 0.00	< 0.01	0.01 ± 0.00	0.11 ± 0.00	< 0.01

Table 7.2. Daily consumption (mean ± SE) of various vegetables in the Akumadan
population (n = 130).

Vegetables	Daily consumption (g/person, fresh weight)	Percentage of total consumption (%)
Tomato	200 ± 47	35
Cabbage	5 ± 2	1
Pepper	10 ± 6	2
Onion	150 ± 92	27
Egg plants	200 ± 38	35
Total	565 ± 108	100

There was a gender-specific difference in the rate of vegetable consumption: women consumed more vegetables (580 g/day) than men (555 g/day). In this survey, five vegetable crops were examined for individual dietary consumption (Table 7.2). Tomato and eggplants, together, accounted for about 70% of total vegetable consumption in the survey population, and the order of consumption was tomato = eggplants > onion > pepper > cabbage.

The results of an evaluation of the non-cancer risks to human health associated with the consumption of vegetables containing DDTs, total endosulfan, HCHs, methoxychlor, dimethoate, and lambda cyhalothrin is shown in Figure 7.2. The hazard ratios (HRs) of non-cancer risk (based on ND = 0 and ND = limit of quantification) at Akumadan were all < 1 for vegetables. The estimated pesticides dietary intakes and their associated HRs were insensitive to the treatment of non-detect residue samples (Table 7.3; Figure 7.2). For instance, changing the non-detect value from zero to the limit of quantification did not markedly change the exposure levels for all of the contaminants. The value assigned to non-detects also did not have an effect on the calculated HRs. The potential health risk of a daily intake of HCHs-contaminated vegetable had a HR of over 0.10 which was due to consumption of tomato (70%) and onion (19%), followed by methoxychlor and dimethoate, with HRs of over 0.05 and 0.03 for vegetables. Onions were found to be the largest contributor to lambda cyhalothrin exposures, accounting for 89% of the estimated distribution. Among the vegetable items, tomato and onion were the principal contributors to the average pesticide exposure among individuals.

Table 7.3. Calculated dietary intake of pesticides from consumption of vegetables

Pesticide components	Daily intake [a] (pg/kg body weight)					
	Tomato	Cabbage	Pepper	Onion	Eggplants	Total intake
ND [b] = 0						
DDTs [c]	89	34	-	45	30	197
Total endosulfan [d]	625	225	-	4129	-	4980
HCHs [e]	655	46	51	179	-	930
Methoxychlor	83958	63	1106	104799	2976	192903
Dimethoate	30	32	-	3929	655	4645
Lambda cyhalothrin	30	-	1	246	-	277
ND = LOQ [f]						
DDTs	89	34	1	45	30	199
Total endosulfan	625	225	1	4129	30	5011
HCHs	655	46	51	179	30	960
Methoxychlor	83958	63	1106	104799	2976	192903
Dimethoate	30	32	1	3929	655	4647
Lambda cyhalothrin	30	1	1	246	30	307

[a] Because of low values, these results are expressed as picogram per kilogram body weight;
[b] Non-detect (that is, below the limit of quantification); [c] *p,p'*-DDE + *p,p'*-DDT; [d] α-endosulfan + β-endosulfan + endosulfan sulphate; [e] α-HCH + β-HCH + γ-HCH; [f] Limit of quantification

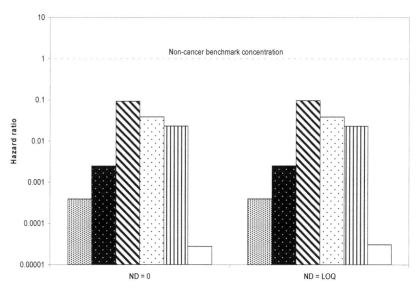

Figure 7.2. Non-Cancer hazard ratios for daily tomato, cabbage, pepper, onion and eggplants consumption by people of Ghana

Discussion

This study examined contamination of vegetable crops sampled across Ghana. Because sampling was performed at central markets across the country, and no remarkable systematic differences were observed between these markets, the results provide a fair picture of the contaminant exposure profile for the general population of Ghana.

In the study, organochlorine (DDTs, endosulfan, HCHs, methoxychlor), organophosphorus (dimethoate), and synthetic pyrethroid (lambda cyhalothrin) pesticide residues were detected in quantifiable amounts in vegetable crops.

According to Trapp *et al.* (1994) and Miglioranza *et al.* (1999), the anatomical and physiological features that enable plants to accumulate nutrients, water, and carbon dioxide also allow them to take up anthropogenic chemicals from air, water and soil via roots (and translocated by the xylem) and, to some extent, across the surface of leaves and other tissues. Several researchers have studied the occurrence of pesticide residues in crops. Gonzalez *et al.* (2003) studied the occurrence and distribution of organochlorine pesticides in tomato crops from organic production and found that endosulfan, HCHs, and DDTs levels in the fruit (edible tissue) were significantly below the maximum residue limits considered by the Codigo Alimentario Argentino and the Codex Alimentarius. Raha *et al.* (1993) measured residues of endosulfan, decamethrin, and fenvalerate in eggplant fruits at concentrations below the tolerance level of 2 mg/kg (for all three insecticides) as specified by FAO/WHO and the Central Insecticide Board of India, and Antonious *et al.* (1998) found 0.95 µg/g endosulfan residues in pepper. Aguilera-del Real *et al.* (1997) reported levels of endosulfan residues in peppers, cucumbers, and cherry tomatoes as 0.16, 0.12, and 0.30 mg/kg, respectively. Monitoring studies conducted in India (Battu *et al.*, 2005) revealed widespread contamination of vegetables, fruits, and cereals with insecticide residues. In vegetables from Eastern Romania (Hura *et al.*, 1999), the mean levels of HCHs and DDTs varied between 0.24 and 1.19 µg/kg in carrots and lettuce for HCHs, and 0.89 and 9.8 µg/kg in carrots, lettuce and cabbage for DDTs. Amoah *et al.* (2006) measured concentrations of *p,p'*-DDT, endosulfan, and lambda cyhalothrin in lettuce in the ranges 0.02-0.9, 0.04-1.3, and 0.01-1.4 mg/kg, respectively. Compared with these data, our results for pesticide residues in vegetables are lower. However, given that vegetables are widely consumed in Ghanaian diet, these results are of great relevance, and deserve closer study.

An important finding of the study is the simultaneous presence of α-HCH and β-HCH together with γ-HCH-the only HCH isomer with insecticide property. In contrast to β-HCH, which is generally abundant among HCH isomers in human (fat, blood, and breast milk) and wildlife (Willett *et al.*, 1998), γ-HCH is the most prevalent isomer in environmental samples. The ratio of γ-HCH to β-HCH determined in this study is about 2:1. Other studies (Willett *et al.*, 1998; Ntow, 2001; 2005; Amoah *et al.*, 2006) have reported the presence of γ-HCH in plant, water, and sediment. Because of the high persistence and bioaccumulation of α-HCH and β-HCH, technical HCH mixtures were banned in almost all developed countries more than 25 years ago, but application of γ-HCH (lindane) is still allowed for specific purposes (Muntean *et al.*, 2003). In Ghana, lindane has been in the provisional list of severely restricted pesticides since 1996 (Gerken *et al.*, 2001). The elevated levels of α-HCH and β-HCH suggest that a technical HCH mixture rather than pure γ-HCH has been used in vegetable cultivation in Ghana in the years preceding the restriction on γ-HCH use.

Although technical DDT is in the Provisional list of Banned Pesticides in Ghana, the Act (Act 528) establishing this ban (which started operation in April 1999) is not fully implemented. This, therefore, leaves room for non-compliance by pesticide dealers. DDT has been used for malaria control programs. The stock of obsolete DDTs may still be illegally distributed and used. Results of the present study revealed that p,p'-DDE occurred in greater concentration in vegetables than p,p'-DDT with a $p'p$-DDE: $p'p$-DDT ratio of about 2:1. p,p'-DDE, which occurs as the main DDT residue in samples, is a metabolite of p,p'-DDT (Jiang *et al.*, 2005), and generally, more persistent in the environment than the parent DDT. Thus, when the use of DDT in a country ceases, the levels of this compound decrease more rapidly than the levels of DDE, thereby producing an increasing DDE/DDT ratio. Interestingly, the p,p'-DDE to p,p'-DDT ratio, which may indicate whether the exposure was in the distant (high ratio) or recent (low ratio) past (Carreño *et al.*, 2006), was estimated at 2 in the present study. This ratio suggests the influence of the prohibition of DDT and the decrease in exposure to these compounds over the years.

It can be hypothesised that crop exposure to pesticides could be from contact with the chemicals during spraying in the vegetable field. Endosulfan, dimethoate, and lambda cyhalothrin are among the most commonly used pesticides in vegetable production in some parts of Ghana. For instance, vegetable farmers in a survey by Ntow *et al.* (2006) admitted to spraying lambda cyhalothrin (Karate 2.5 EC/ULV), dimethoate (Perferkthion 400 EC) and endosulfan (Thionex 35 EC/ULV, Thiodan 50 EC) on tomato, pepper, okra, eggplant, cabbage and lettuce. Although DDTs, HCHs and methoxychlor did not appear to be used in vegetable farming (they are banned from, or restricted in, agricultural use), their occurrence in the samples is not surprising. This is mainly because these chemicals (organochlorines), relative to other classes of pesticides are resistant to environmental degradation, which allows them to accumulate in plant and animal tissue. Besides, vegetable farmers in Ghana use pesticides which are not registered and which may have found their way into the country though unapproved routes. Generally, vegetable farming in Ghana is fraught with abuse, misuse and overuse of pesticides. Pests and diseases pose big problems in vegetable production and these have led many farmers to use chemical pesticides, even if they have received no training in the choice of chemicals or application technique. The results regarding pesticide residues also indicated that several pesticides are used within a crop-growing season. As also described by Danso *et al.* (2002) and Ntow *et al.* (2006), vegetable farmers mix cocktails of various pesticides to increase the potency of the compounds. Further investigations are ongoing to clarify pesticide exposure pathways.

Since some vegetables, like those considered in this study, are traditionally very important for "stews" and soups, which accompany the main national dishes like the so called "ampesi" and "fufu", respectively, in Ghana, intake of especially toxic pesticides from vegetables is of great concern for human health risk. This study provided insight into the magnitude of potential exposures from vegetable contamination. Initial screening of the data presented showed that estimated exposure to a number of contaminants in the average vegetable diet of adults did not exceed benchmark concentration for non-cancer effects. This indicates that exposure to pesticides was unlikely to result in any adverse health effects. Dietary intakes of over 90 analytes presented for eight population groups in the United States revealed average daily intakes well below acceptable limits (Gunderson, 1995). Gold *et al.* (1997), using standard methodology, measured dietary residues in the total diet and reported the estimate of excess cancer risk from average lifetime exposure to synthetic pesticide residues in the diet to be less than one in a million for each of the 10 pesticides analysed. In a duplicate study, during 1987-1991 in Bavarian homes (Bavaria, Germany) for young and elderly people, pesticides were detected in 15% of the samples. The pesticide content reached 8% of the respective FAO/WHO limits (Arnold *et al.*, 1998). However, it is important to note that some of the identified pesticides in this study, for example DDT and its metabolite DDE, together with β-HCH are known to suppress lactation (Muntean *et al.*, 2003). These compounds are highly persistent and their continued presence at significant levels in the food supply is a matter of great health concern. Exposure to these "environmental estrogens" leads to a reduction in the amount of breast milk and fat content and is potentially harmful to infants, especially in deprived conditions (Perez-Escamilla, 1993; Amador *et al.*, 1994). This study suggests that concerted efforts should be made by the regulatory authorities in Ghana for the safe and judicious use of agrochemicals for pest control to reduce body burdens of these chemicals in foodstuffs to safe levels.

Health risk assessment is very subjective (Raschke and Burger, 1997) and is only as accurate as the available information. There are a number of important limitations in this study. A first source of uncertainty in this analysis arises from the vegetable consumption data. We collected vegetable consumption data from a sub-population of farmers at Akumadan, which may not be representative of dietary patterns of the entire Ghanaian population. The consumption data obtained in this survey were higher than data presented by Ghana Living Standard Survey (http://www.worldbank.org/html/prdph/lsms/country/gh/docs/G4report.pdf; accessed on March 14, 2007). This is not surprising since our sample population consist of a community in which a large proportion of the food consumed comes from their own produce. At the same time, such dietary exposure may only begin to describe the total body burden on local residents consuming contaminated water sources and breathing air containing contaminated dust. Other equally important uncertainties, which the present study did not consider, are cancer risks, possible interactions among various toxic chemicals, and different vegetable consumption patterns within and between populations. Overall, despite the limitations associated with the analysis, the results point to potential exposures to contaminants in vegetables and represent an important step toward better characterisation of these exposures.

Acknowledgements

The work described in this paper was supported by a research grant from the Dutch Government (through the UNESCO-IHE Institute for Water Education, Delft), International Water Management Institute, and from the International Foundation for Science. The authors thank Augustina Kumapley, Rafia Mamudu, Sarah Kafui, Ato (Akumadan) and vegetable farmers in Ghana for their assistance and cooperation. Sampson Abu and Bernice Worlanyo Ntow are also acknowledged for their technical and secretarial assistance, respectively. The Kinneret Limnological Laboratory, Migdal, Israel, is acknowledged for technical assistance in the use of GC/MS.

References

Aguilera-del Real A, Valverde-García A, Fernandez-Alba AR, Camacho-Ferre F (1997) Behaviour of endosulfan residues in peppers, cucumbers and cherry tomatoes grown in Greenhouse: Evaluation by decline curves. Pestic Sci 51:194-200

Amador M, Silva LC, Valdés-Lazo FSO (1994) Breast-feeding trends in Cuba and the Americas. Bull Pan Am Health Organ 28(3):220-228

Amoah P, Drechsel P, Abaidoo RC, Ntow WJ (2006) Pesticide and Pathogen Contamination of vegetables in Ghana's Urban Markets. Arch Environ Contam Toxicol 50:1-6

Antonious GF, Byers ME, Snyder JC (1998) Residues and fate of endosulfan on field-grown pepper and tomato. Pestic Sci 54:61-67

Arnold R, Kibler R, Brunner B (1998) Alimentary intake of selected pollutants and nitrate-results of a duplicate study in Bavarian homes for youth and elderly people. Z Ernahrungswissenschaft 37:328-335.

Battu RS, Singh B, Kang BK, Joia BS (2005) Risk assessment through dietary intake of total diet contaminated with pesticide residues in Punjab, India, 1999-2002. Ecotoxicology and Environmental Safety 62:132-139

Bull D (1982) A growing problem: Pesticides and the Third World poor, OXFAM, Oxford, 192 pp

Carreño J, Rivas A, Granada A, Lopez-Espinosa MJ, Mariscal M, Olea N, Olea-Serrano F (2006) Exposure of young men to organochlorine pesticides in Southern Spain. Environmental Research doi: 10.1016/j.envres.2006.06.007.

Danso G, Drechsel P, Fialor SC (2002) Perception of organic agriculture by urban vegetable farmers and consumers in Ghana. Urban Agric Mag 6:23-24 Dinham B (2003) Growing vegetables in developing countries for local urban populations and export markets: problems confronting small-scale producers. Pest Manag Sci 59:575-582

Dougherty CP, Holtz, SH, Reinert JC, Panyacosit L, Axelrad DA, Woodruff TJ (2000) Dietary Exposures to Food Contaminants across the United States. Environmental Research Section A 84:170-185

Ferrer I, García-Reyes JF, Mezcua M, Thurman EM, Fernández-Alba AR (2005) Multi-residue pesticide analysis in fruits and vegetables by liquid chromatography-time-of–flight mass spectrometry. Journal of Chromatography A 1082:81-90

Gerken A, Suglo JV, Braun M (2001) Pesticide policy in Ghana. MoFA/PPRSD, ICP Project, Pesticide Policy Project/GTZ, Accra, 2001, pp 185

Gold LS, Stern BR, Slone TH, Brown JP, Manley NB, Ames BN (1997) Pesticides residues in food: investigation of disparities in cancer risk estimates. Cancer Lett 195-207

Gonzalez M, Miglioranza KSB, Aizpún de Moreno JE, Moreno VJ (2003) Occurrence and distribution of organochlorine pesticides (OCPs) in tomato (*Lycopersicum esculentum*) crops from organic production. J Agric Food Chem 51:1353-1359

Gunderson EL 91995) Dietary intake of pesticides, selected elements and other chemicals: FDA total diet study, June 1984-April 1986. JAOAC Int 78:910-921

Hsu R-C, Biggs I, Saini NK (1991) Solid-Phase Extraction Cleanup of Halogenated Organic Pesticides. J Agr Food Chem 39:1658-66

Hura C, Leanca M, Rusu L, Hura BA (1999) Risk assessment of pollution with pesticides in food in the Eastern Romania area (1996-1997). Toxicology Letters 107:103-107

Jiang Q, Hanari N, Miyake Y, Okazawa T, Lau RKF, Chen K, Wyrzykowska B, So MK, Yamashita N, Lam PKS (2006) Health risk assessment for polychlorinated biphenyls, polychlorinated dibenzo-*p*-dioxins and dibenzofurans, and polychlorinated naphthalenes in seafood from Guangzhou and Zhoushan, China. Environmental Pollution doi: 10, 1016/j.envpol.2006.11.002,1-9

Kroes R, Müller D, Lambe J, Löwik MRH, van Klaveren J, Kleiner J, Massey R, Mayer S, Urieta I, Verger P, Visconti A (2002) Assessment of intake from the diet. Food and Chemical Toxicology 40:327-385

MacIntosh DL, Spengler JD, Ozkaynak H, Tsai L, Ryan PB (1996) Dietary exposures to selected metals and pesticides. Environ Health Perspect 104:202-209

Miglioranza KSB, Aizpún de Moreno JE, Moreno VJ, Osterrieth ML, Escalante AH (1999) Fate of organochlorine pesticides in soils and terrestrial biota of "Los Padres" pond watershed, Argentina. Environ Pollut 105:91-99

Muntean N, Jermini M, Small I, Falzon D, Fürst P, Migliorati G, Scortichini G, Forti AF, Anklam E, von Holst C, Niyazmatov B, Bahkridinov S, Aertgeerts R, Bertollini R, Tirado C, Kolb A (2003) Assessment of Dietary Exposure to Some Persistent Organic Pollutants in the Republic of Karakalpakstan of Uzbekistan. Environ Health Perspect 111(10):1306-1311

Ntow WJ (2001) Organochlorine pesticides in water, sediment, crops, and human fluids in a farming community in Ghana. Arch Environ Contam Toxicol 40:557-563

Ntow WJ (2005) Pesticide residues in Volta Lake, Ghana. Lakes & Reservoirs: Research and Management 10:243-248

Ntow WJ, Gijzen HJ, Kelderman P, Drechsel P (2006) Farmer perceptions and pesticide use practices in vegetable production in Ghana. Pest Manag Sci 62(4):356-365

Perez-Escamilla R (1993) Breast-feeding patterns in nine Latin American and Caribbean countries. Bull Pan Am Health Organ 27(1):32-42

Raha P, Banerjee H, Das AK, Adtyachaudhury N (1993) Persistence kinetics of endosulfan, fenvalerate, and decamethrin in and on eggplant (*Solanum melongena* L.). J Agric Food Chem 41:923-928

Raschke AM, Burger AEC (1997) Risk assessment as a management tool used to assess the effect of pesticide use in an irrigation system, situated in a Semi-Desert Region. Arch Environ Contam Toxicol 32:42-49

Trapp S, Mc Farlane C, Matthies M (1994) Model for uptake of xenobiotic into plants: validation with bromacil experiments. Environ Toxicol Chem 13:413-422

USEPA (1989) Exposure factors handbook. Exposure assessment group. Office of Health and Environmental Assessment. U.S Environmental Protection Agency, Washington, DC

Willett KL, Ulrich EM, Hites RA (1998) Differential Toxicity and Environmental Fates of Hexachlorocyclohexane Isomers. Environ Sci Technol 32(15):2197-2207.

Chapter 8

Conclusions and Recommendations

Conclusions and Recommendations

This Thesis has examined and compiled the results of a scientific study to identify the practices of use of pesticides in vegetable cultivation in Ghana, and the incidence and magnitude of their on-site and off-site environmental effects. The knowledge of human exposure to and potential public health risks of levels of current-use pesticide residues has been discussed. In addition, the Thesis has identified the data and knowledge gaps that need to be filled by future pesticide residues research and monitoring. The Thesis drew on existing knowledge and incorporated new information from the research funded under the UNESCO-IHE/IWMI/IFS funding programme. The most important conclusions drawn from the study, alongside recommendations, will be presented here.

Profile of vegetable pesticide usage
Vegetable farmers in Ghana have been applying pesticides for over four decades. Most farmers, particularly those at Akumadan, make two to twelve applications of pesticide in a typical growing season of tomatoes, pepper, eggplants or okra. The pesticides used are organophosphates such as chlorpyrifos and dimethoate, as well as organochlorines such as endosulfan, and pyrethroids such as lambda cyhalothrin, cypermethrin and deltamethrin. Of the 43 chemicals used on sample farms in this Research, only four are registered for general use in Ghana. The others are meant for restricted use or are not registered for use at all. Unsafe pesticide storage, handling, and disposal practices coupled with inadequate education and training, documented in Chapter 2, subject the farmer to high levels of health hazards and contaminate the vegetable ecosystem. Clearly, there is an urgent requirement for well-targeted training programmes for vegetable farmers on the need for and safe use of pesticides.

On-site and off-site environmental effects of vegetable pesticides
In Chapter 3, field trials were used to understand the fate of a typical pesticide on a vegetable plot. The trial experiments were conducted in a field at Akumadan, the most prominent vegetable cultivating community in Ghana. The field trials, together with the survey of Chapter 2, provided a means to identify "hotspots" of environmental effects and pathways of human exposure of pesticides. The pesticide and vegetable used were endosulfan and tomato (*Lycopersicum esculentum*), respectively. The selection of the pesticide was based on use, stability (*i.e.* biodegradability), toxicity and bio-accumulative properties. Tomato was selected for this study because it is widely grown in the study area, related to the use of the selected pesticide and due to the economic importance of its crop.

Despite the loss of most of the endosulfan applied onto field-grown tomato at Akumadan, residues were found in plant material, in the few hours after treatment, in concentrations sufficient to pose a risk if tomato crops were consumed as fresh fruit.

Therefore, it is preferable to allow at least a 2-week pre-harvest interval during which no application of pesticide is carried out, to allow for the dilution of the chemical to take place with the growth of the fruit. After such an interval, the residue concentration will have decreased to a level below the Codex[1] MRL.

Tomato fields can act as strong sources of pesticide residues in runoff water for several months after application; apart from parent isomers of endosulfan, the residues also include the equally toxic sulfate metabolite, which persists for several months in soil. In Chapter 4, there is clear evidence that in both study areas, Akumadan and Tono, small streams draining vegetable farmlands receive relatively high loadings of endosulfan and chlorpyrifos used by farmers for crop protection in runoff water. Although seasonal mean concentrations for endosulfan and chlorpyrifos in water-phase and streambed sediment were found to be below threshold levels for the safety of humans and aquatic invertebrates, a major concern lies in the additive and, sometimes even synergistic, action of multiple pollutants. Consequently, irrigation and storm runoff must, as much as possible, be retained on-farm by proper management of water, including the use of drainage ditches and the building of large water storages. As another possible strategy, the implementation of constructed wetlands is recommended. Nevertheless, in very large storms it may be impossible to prevent movement of endosulfan and chlorpyrifos residues off-farm to nearby streams and rivers. In this sense, regular monitoring of the quality of freshwater resources is recommended. In Ghana, there is no clear policy for pesticide concentrations in effluents from various sources discharging into freshwater bodies. Without such standards, it will be impossible to regulate the water quality of such water bodies. In Chapter 2 of this Thesis, it was found that endosulfan residues did not leach beyond 20 cm depth in sandy loam soil.

Human exposure and potential public health risks: blood enzyme inhibition
At Akumadan, about 70% of all exposed farmers had blood ChE activity levels at or below 70% of the normal reference mean (that is, a reduction of 30% or more in ChE activity). This inhibition of enzyme activity could be due to high exposure to pesticides because of lack of safe spray equipment and protective clothing suitable for tropical conditions (previously reported in Chapter 2). Pesticides that might be linked with these inhibitions include certain organophosphates and carbamates. The conclusion that was reached in Chapter 5 was that organophosphates were the anti-ChE pesticides responsible for the blood ChE inhibition in farmers at Akumadan. However, literature information suggests that potentially anti-ChE pesticides are not likely to be limited to any particular class, citing other chemicals, including other pesticides, and detergents as potential anti-ChE agents. Therefore, future research must be directed at obtaining information on exposure to chlorinated hydrocarbons, detergents, as well as metals such as cadmium, copper, mercury, lead and uranium.

Combined use of pesticides was found to be common in vegetable production (Chapter 2). Although total exposure to combined organophosphates or combined carbamates can be assessed with the use of ChE monitoring, there is a lack of sufficient knowledge on the interaction of different pesticides in the human body during combined exposure. Research on the biological monitoring of occupational exposure to combined pesticides is recommended. In addition, it is recommended to assay urinary metabolites of pesticides in occupationally exposed subjects and in the general population of Ghana, since these metabolites are much more sensitive indicators of exposure than cholinesterase activity.

The assessment of exposure and health risks detailed in Chapter 5 was done by comparing the levels of ChE of the exposed group to that of a control group. The levels of ChE determined for the control group served as baseline or reference values for comparison of the health effects of pesticide exposure. However, to make the interpretation of results more meaningful, it is recommended to carry out research to compare the farmers' post-exposure ChE level with their own baseline value.

It is more difficult to conduct biological monitoring in farmers who handle pesticides in the field. From a practical point of view, it is recommended to promote research of biological monitoring in routine use for making sampling acceptable, analytical methods applicable, and sources of variability avoidable.

Human exposure and potential public health risks: residues in body fluids
It appeared from the results of the study reported in Chapter 6 that the concentrations of pollutants in human samples from Ghana are situated at the low end of the concentration range measured in many countries. However, the presence of HCHs in human breast milk was an important finding. In view of this, I suggest that further investigations on the pollution and sources of HCHs in Ghana are needed to assess human health risks. Moreover, because organochlorine compounds can be mobilised during pregnancy and lactation, the assessment should focus initially on pregnant women and newborn children because they are the most vulnerable to these toxic substances.

Human exposure and potential public health risks: food contamination
The study reported in Chapter 7 pioneers the collection of farmers' vegetable consumption data to establish the Ghanaian dietary exposure to pesticide residues. Vegetable dietary exposure to pesticide residues is low and there is no associated health risk. However, the results of the study indicated that contaminated vegetables are the main pesticide exposure route. The urgent need now is for efforts to ensure that the public is better informed about the risks from contaminants in foodstuffs and, in particular, how they can reduce their personal exposure by simple procedures established in literature, such as adequate washing, peeling, and cooking. Scientific efforts to characterise exposure from drinking water are also needed, as contamination of drinking water by pesticides represents a growing problem in many areas of the country, particularly in agricultural areas.

Concluding remark
The concluding remark of this study is that pesticides used in Ghanaian vegetable production are associated with water pollution, food contamination, ChE inhibition, and accumulation of toxic compounds in human fluids for farmers. In many respects, Akumadan presents a worst-case example of the problems of the use of pesticides in vegetable production. Because pesticide use is such a major component in the environmental and human health issues described in this Thesis, its reduction-through both individual and collective action-can have profound effects on the health of humans and the environment. The management and control of the environmental and health effects of pesticides should be based on a simple principle: risk is determined by exposure, but the best predictor of exposure is use. Therefore, a reduction in use should lead to a reduction of exposure and risk. Throughout the world, environmental abuse and health risks result from excessive use of pesticides. Literature information shows that increased pesticide use leads to increased pest resistance, leading to even more use, resulting in adverse effects on human health and in an often devastating impact on the ecosystem. Drawing on nations' experiences from literature, I conclude this Thesis by proposing effective actions such as pesticide substitutions (for example, neem extracts), biological control, and integrated pest management as solutions of this problem for Ghana.

Note

1. The Codex Alimentarius (Latin for "food code" or "food book") is a collection of internationally recognised standards, codes of practice,guidelines and other recommendations relating to foods, food production and food safety under the aegis of consumer protection. These texts are developed and maintained by the Codex Alimentarius Commission, a body that was established in 1963 by the Food and Agriculture Organisation of the United Nations (FAO), and the World Health Organisation (WHO). The Commission's maim aims are stated as being to protect the health of consumers and ensure fair practices in the international food trade. The Codex Alimentarius is recognised by the World Trade Organisation as an international reference point for the resolution of disputes concerning food safety and consumer protection (http://en.wikipedia.org/wiki/Codex_Alimentarius; accessed on March 25, 2007).

Samenvatting

Zoals in andere tropische landen is de economie van Ghana gestoeld op landbouw en daardoor afhankelijk van een intensief gebruik van pesticiden voor productieverbetering. Pesticiden in het milieu vormen een probleem vanwege de kans op vervuiling van voedselproducten en kwetsbare ecosystemen zoals rivieren en beken. In Ghana bestaat een toenemende kans op nadelige gezondheids- en milieueffecten, vooral omdat maatregelen om deze effecten te minimaliseren, vaak niet worden toegepast.

In 2003 werd met dit Promotieonderzoek begonnen, dat zich richtte op het identificeren en kwantificeren van locale en niet-locale milieueffecten van het gebruik van pesticiden in de productie van groenten. In de afzonderlijke hoofdstukken van dit Proefschrift worden de methodologieën en resultaten van de verschillende componenten van het onderzoek uitvoerig besproken.De locale milieueffecten werden bestudeerd in veldexperimenten op semi-praktijkschaal met een tomatencultuur; hierbij werd gebruik gemaakt van endosulfan dat als Thiodan 35 EC/UVL op de tomatenplanten werd gesproeid. In deze experimenten werd de dissipatie en persistentie van endosulfan in de grond, de tomatenplanten en de vruchten onderzocht. Ook werd onderzoek gedaan naar de trends in vervuiling van het afvoerende water. Met deze vervuiling is er vaak ook spake van blootstelling van de lokale boerenbevolking; dit werd epidemiologisch onderzocht door het meten van de activiteit van het enzym cholinesterase. Het bleek dat in de praktijk de mate van blootstelling afhankelijk was van de mate van bescherming door de boeren bij het toepassen van de pesticiden.

Residuen van α- and β-endosulfan en van endosulfan-sulfaat werden in de grond van de tomatencultuur tot enkele dagen na toepassing gedetecteerd; daarentegen werd het meer moeilijk-afbreekbare, en even giftige, metaboliet endosulfaan-sulfaat nog na enkele maanden in de bodem aangetroffen. De residuewaarden van endosulfan in de tomatenvruchten zelf lagen steeds onder de Codex MRL standaardwaarden. Gedurende de perioden van hogere waterafvoeren werden tijdelijk verhoogde residuewaarden van de gebruikte pesticiden gevonden in het afvoerende water benedenstrooms van de groentenakkers. De primaire blootstelling met betrekking tot de verschillende organochloor-, organofosfaat- en pyrethroïdenverbindingen vond plaats door inademing, via de huid en door het traditionele groentengebruik. Dit geldt met name voor de boeren, door het niet gebruiken van beschermende kleding en andere onjuiste praktijken tijdens het, vooral, handmatig sproeien. Ca. 70% van de blootgestelde boeren had een 30% of hogere afname in de cholinesterase-activiteit in het bloed. Rond 95% had klachten over hoofdpijn en algemene verzwakking. Dit is echter niet in overeenstemming met hun gemeten lage dagelijkse pesticideninname via voedsel. In het bloedserum en de moedermelk van de boerenbevolking werden ook residuen gevonden van de moeilijk-afbreekbare pesticiden DDT, dieldrin, HCB en HCH. Gezien de gemeten dagelijkse inname van DDT en HCH via borstvoeding bestaat er hier reden tot zorg voor de gezondheid van kinderen.

About the author

 William Joseph Ntow, born in February 1960 in Kumasi, Ghana, had his secondary education at Opoku Ware School, Kumasi. His university education began at the University of Cape Coast, Ghana, where he obtained an honours degree in Chemistry and a concurrent Diploma in Education in 1986. He taught secondary school chemistry for four years and proceeded to the University of Science and Technology (now Kwame Nkrumah University of Science and Technology), Kumasi, Ghana in October 1990 to pursue a two-year Master's programme in Environmental Chemistry. His research career began at the Institute of Aquatic Biology (now Water Research Institute, WRI) of the Council for Scientific and Industrial Research (CSIR) as an Assistant Research Officer (now Assistant Research Scientist) in September 1993. In February 1994, he was upgraded to a Research Officer (Research Scientist), having submitted an MPhil certificate from KNUST. Between the period 1994 and October 1996, he carried out research into various aspects of the resources of inland, estuarine, lagoonal and the immediate coastal inshore water systems of Ghana covering water quality and pollution. He wrote several reports and spoke at several conferences and seminars. In October 1996, he won a Netherlands Government Scholarship to study Water and Environmental Resources Management at the International Institute for Infrastructural, Hydraulic and Environmental Engineering (now UNESCO-IHE Institute for Water Education) in Delft, The Netherlands. He further studied Limnology at the Institute of Limnology (Austrian Academy of Sciences), Mondsee, Austria, and the Institute of Botany and the Institute of Parasitology (all at the Czech Republic Academy of Sciences), Czech Republic. He obtained an MSc at UNESCO-IHE in 1998, specialising in Water Quality Management.

Ntow still works for the CSIR WRI as a Research Scientist. His research interests are related to the assessment of water quality in relation to pollution from agricultural activities. He has several publications in his name in peer-reviewed journals. Ntow was appointed Project Manager of DANIDA/Ghana Government project on 'Strengthening of Water Resources Information Services' in 1999. He was the Head of the Environmental Chemistry Division of the CSIR WRI between 1999 and 2002. He has undertaken several consultancy works related to Environmental Impact Assessment. He is a member of the Ghana Science Association, Research Staff Association of the CSIR, and the Ghana Chemical Society.

Ntow is married to Bernice Worlanyo Ntow (Mrs) and has three children studying in the United States of America. He is a member of the Word Miracle Church International and committed to the Christian faith.

Supervision, Training and Education Plan (STEP) – Sandwich model PhD

PhD Seminars, UNESCO-IHE, Delft
- 2003 (attended)
- 2004 (attended and made a presentation)
- 2005 (attended and made a presentation)

Role-play, UNESCO-IHE, Delft
- 2003 (participated)
- 2004 (participated)
- 2005 (participated)

UNESCO-IHE Lunch Seminar
- 2004 (attended and made a presentation)

International Congresses/Workshops
- 2005 (IFS Workshop, Ouagadougou)
- 2005 (BCPC Congress, Glasgow)
- 2006 (IFS/OPCW Workshop, Nairobi)

Training of Trainers
- 2004 (MSc supervision, Phenny Mwaanga, Zambia)
- 2004-2006 (MPhil supervision)
  1. Laud M. Tagoe, Ghana
  2. Benjamin O. Botwe, Ghana
  3. Jonathan Ameyibor, Ghana

RegularUNESCO-IHE MSc Courses
- 2003 (Geographic Information Systems)
- 2004 (Poster Presentation)
- 2005 (Thesis Writing)

Specialised Courses Elsewhere
- 2003 (GC-MS, KLL, Israel)
- 2005 (GC-MS, TUDelft, Delft)

T - #0114 - 071024 - C12 - 254/178/7 - PB - 9780415462747 - Gloss Lamination